Industrial Image and Process in Contemporary Art

painting MACHINES

Boston University Art Gallery
855 Commonwealth Avenue
Boston, Massachusetts 02215

Distributed by the University of Washington Press
P.O. Box 50096, Seattle, Washington 98145

Printed in the United States of America
Library of Congress Catalogue Card Number: 96-084502
ISBN: 1-881450-07-4

Cover Illustrations:
(front) Donald Sultan, *Plant, May 24, 1985*, 1985.
(back) Rosemarie Trockel, *Untitled (Woolmark)*, 1986; Mark Tansey, *The Raw and the Framed*, 1992.
(both) Angela Bulloch, *Pushmepullme Drawing Machine*, 1991.

painting MACHINES

Industrial Image and Process in Contemporary Art

Exhibition and catalogue by Caroline A. Jones

Exhibition coordinated by John R. Stomberg

Boston University Art Gallery

October 30–December 14, 1997

University of Washington Press ❒ Seattle and London

CONTENTS

ACKNOWLEDGMENTS

Painting Machines pursues the Boston University Art Gallery's ongoing investigation of the machine as a visual and cultural entity with enduring meaning in the United States and abroad. It is a logical outgrowth of several recent projects at Boston University and the Art Gallery: in a series of interdisciplinary conferences and exhibitions, the University has encouraged the intellectual exploration of twentieth-century attitudes toward technology. For example, the "Histories of Science, Histories of Art" symposium in November 1995 was jointly sponsored by Boston University and Harvard University, and was co-organized by Professor Caroline Jones, the curator of *Painting Machines*. The lectures and discussions were accompanied by an installation of transformed household machinery, *Faraday's Islands*, by artist Perry Hoberman. In 1995 we also organized the exhibition *From Icon to Irony: German and American Industrial Photography,* which juxtaposed heroic photographers of 1920s industry (Margaret Bourke-White, Germaine Krull, Albert Renger-Patzsch, Charles Sheeler) with contemporary critiques of the technological world (Bernd and Hilla Becher, Joachim Brohm, Frank Gohlke, John Pfahl).

Painting Machines would not have been possible without the generous contributions and support of many individuals and institutions. We are grateful to Professor Katherine T. O'Connor and the Humanities Foundation of the College of Arts and Sciences, and Graduate School, Boston University, for their support of this catalogue. We also thank Jürgen Keil, Beeke Sell Tower, and the Goethe-Institut Boston for their assistance with this project. Additionally we thank Phyllis Rosenzweig, Associate Curator, and Margaret Dong, Assistant Registrar, Hirshhorn Museum and Sculpture Garden, Smithsonian Institution; Stuart Regen, Shaun Caley and Suzanne Unrein, Regen Projects, Los Angeles; Brian D. Butler, 1301 Projects and Editions, Santa Monica; Bruce Wick and Paris Murray, Joseph Helman Gallery; Monika Sprüth, Monika Sprüth Galerie, Cologne, Germany; Graham W. J. Beal, Director and Executive Vice President, Stephanie Barron, Senior Curator of Twentieth-Century Art, and Susan Oshima, Assistant Registrar, Los Angeles County Museum of Art; Juliana L. Hanner, Registrar at the Eli Broad Family Foundation, Santa Monica, and the Eli and Edythe L. Broad Collection, Los Angeles; Jeffrey Deitch and Simone Manwarring, Deitch Projects, New York; Curt Marcus and Katie Gass, Curt Marcus Gallery, New York; Marjory Jacobson and Marshall Smith, Boston; Ira and Lori Young, West Vancouver, Canada; Paul Barratt, London Projects, London; Marian Goodman, Lane Coburn and Genie Freilich, Marian Goodman Gallery, New York; Pika Keerseblick and Barbara Gladstone, Barbara Gladstone Gallery, New York; Emily Braun, Independent Curator, New York; and the private collectors who have lent works to the exhibition.

We would like to acknowledge the participants of Professor Jones's seminar, from which this exhibition emanates. In particular we thank the essay contributors, Isabelle Sobin, Leslie Goldman, Alice Kim, Ana de Azcárate, Renato Rodrigues da Silva, Karen Gramm, and Anthe Constantinidou. We also thank Ray Garraffa and Kyung-Jin Rhee.

We are grateful for the Art Gallery staff who worked to bring this exhibition together. Lawrence Hyman and Isabelle Sobin assisted with the exhibition from the beginning. We also thank Rachelle A. Dermer, Katie Delmez, Michèle Furst, Karen Georgi, Katrina A. Jones, Alice Kim, and Stacey McCarroll.

Kim Sichel
Director

John R. Stomberg
Assistant Director

Mark Tansey

The Raw and the Framed, 1992

Toner on paper

6 x 10 in.

The Eli and Edythe L. Broad
Collection, Los Angeles

Caroline A. Jones

The painting (of) Machines

mechanolatry

Mechanolatry. This neologism aims to capture, in a word, the centuries-long cult of the machine. It seems that we humans are forever attempting to come to terms with our technological creations; worship is but one response. Our century has had a particularly ambivalent reaction to its machines, alternately looking to them for salvation and fearing them as agents of destruction. The exhibition *Painting Machines*—and this essay—examines our current ways of thinking about the technological, particularly in relation to that most human of activities, making art.

The modest wordplay of the exhibition's title intends to encompass two very different methods of coming to terms with the image of the machine in contemporary culture. Pursuing *painting* as if it were a simple verb, Lawrence Gipe, Robert Moskowitz, Donald Sultan, and Mark Tansey portray machines as entities in the world, as aspects of our visual culture that can be depicted, however ironically, within a painted picture. Sculptors Angela Bulloch, Rebecca Horn, Rosemarie Trockel, and Liz Larner are grouped under the adjectival rubric of the word, making *painting* machines that perform art-making activities. The paintings in the exhibition are landscapes in which the machine has replaced nature; the sculptures are works of performance art conducted through the mediating persona of the machine.

Before we can fully understand these phenomena, however, we need to know the complex history of mechanolatry, a history addressed in this exhibition by both the paintings *of* machines, and the painting *by* machines. Machines occupy a peculiar position in our late-twentieth-century imagination. Since the late 1980s, the machine has been seen increasingly as a crude but necessary vehicle for transporting and mobilizing the fictively disembodied experience of electronics. Given the new power of an emerging digital culture, the machine is seen by contrast as hardware rather than software, constraining body rather than transcending mind—just so much senseless, earthbound "meat" when compared to the cyber-fantasies of an inexhaustible flight through virtual reality.[1] In current discourse, "body" is to "mind" as "machine" is to "cyberspace."

We might date the beginning of this demotion of the machine to the 1960s, and pinpoint one of its most salient moments in New York, at the Museum of Modern Art's 1968 exhibition *The Machine as Seen at the End of the Mechanical Age*. The metal-clad catalogue for this show (fitted with twin piano hinges rather than an organic cloth spine) synoptically surveyed European and North American cultures' centuries-long fascination with things mechanical, concluding with a final section that was printed in a cool, futuristic blue. This epilogue presented documentation of the new synergistic collaborations between art and technology, heralding the beginning of an age of cybernetics and information technology that has largely come to pass.[2] As the Modern's exhibition proclaimed it, the machine was nothing but a clanking anachronism, a rusting relic belonging to another era.

Although *The Machine as Seen at the End of the Mechanical Age* exhibition made it official, the end of the mechanical age had already been acted out at the Museum several years earlier by the Swiss artist Jean Tinguely. Tinguely's relationship with his national heritage of precision clock-making seemed fiercely ambivalent by the time he mounted a kinetic sculpture in the Museum's sculpture garden in the first few weeks of March in 1960; it was titled *Homage to New York,* and appeared to observers as a precarious bricolage of piano interiors, springs, bicycle and baby carriage wheels, horns, bells, fan belts, and rolls of paper. At the appointed hour of 7:30 p.m. on March 17, it was ceremoniously set in motion before an expectant, if not apprehensive, crowd. For twenty minutes it spun, sputtered, groaned, burned, and spewed things, finally collapsing with the aid of a few firefighters, whose axes put an end to it and to the evening's entertainment. As one chronicler put it, "The nineteen sixties have begun."[3] It was a fittingly complex beginning for a decade marked by escalating waves of technophilia and by the emergence of a technophobic cluster of "back to nature" movements.

This demolition and demotion of the machine to a pile of all-too-physical junk seemed to constitute a dramatic change from its status in previous centuries. Beginning with the Enlightenment and extending through the Industrial Revolution and beyond, the machine had enjoyed a long and cerebral reign. It was seen above all as the instantiation of the mind's rational powers of control. In the thinking of Réné Descartes, and even more in the explicitly mechanistic philosophy of *L'homme-machine* (1748), written by his follower Julien Offray de La Mettrie, the machine had been far more than a mere extension of the human *body.*[4] It was seen to express the previously sacrosanct order of the human *soul* (a view Gilbert Ryle later criticized as Descartes's mystical invocation of "the ghost in the machine").[5] The smooth workings of beautifully tooled brass gears and well-oiled regulators became the primary Enlightenment metaphor for a God still secure in His Heaven, a prime mover whose divine impersonal mechanisms ruled the celestial and animal kingdoms with an even, precise, and rational hand.

This smoothly functioning, happy universe was not the only one, however. It coexisted with any number of libidinous parallel worlds. Even Denis Diderot, that author of the Enlightenment par excellence, fantasized about an exotic clime where one might find strange beings possessing a kind of geometric sexuality of bolts and screws, a fantastic island of mechanical "boys' cylindrical, parallelepiped, pyramidal jewels" found alongside girls' equally mechanical "bolt-form" jewels: "circular . . . square . . . [or] irregular-sided." The coupling of these glittering, precious erotic solids would be regulated by male and female thermometers, subsidiary "ingenious machines" (in Diderot's words) that were perfectly designed to mesh with the primary gendered geometries of the bolts and their screws.[6] This vision of an artistic and potentially libidinous technology accompanied many legends of genius. Albertus Magnus was said to have had a robot, and Descartes himself was rumored to have constructed a fantastic mechanical woman (named Francine, no less, after his illegitimate daughter). Accompanying him on long, lonely voyages, she was supposedly discovered in Descartes's luggage by a nosy neighbor and destroyed by the ship's captain, fearful of what seemed to be a product of black magic rather than mechanical genius.[7] (If women were bad luck on a ship, then strangely animated mechanical women were no doubt worse.) Coeval with the rationality of machines, then, came their role as projections of our irrational, desiring, artistic selves.

The Enlightenment is not known for such erotic and magical universes. They lay largely unrecognized, shadowy parallels to the canonical texts' evocation of a divine rationality made manifest in the everyday life of machines. Buried in the *Oeuvres* of Diderot or in gossip about Descartes, such tales proliferated only at the edges of Enlightenment narratives of Progress, marginalized even further as the industrializing engines of the nineteenth century clanked on. In the main, the machine became wedded to a transcendent sublimity that, on the one hand, dwarfed its human observers or, on the other, was seen as a pure expression of human will. Sublimity was the mode sought by painters—most gloriously expressed in J. M. W. Turner's *Rain, Steam, and Speed: The Great Western Railway,* of 1844, or in Claude Monet's dramatic series of interior landscapes painted in the Gare Saint-Lazare, Paris, three decades later.[8] Elsewhere in Western culture the machine was domesticated, appearing as gaslight in the parlor and steam-irons in the laundry-room. It even became an increasingly popular therapeutic tool, as electrical apparatuses were applied directly to the bodies of those suffering from "nervous exhaustion."[9]

In this nineteenth-century frame, the darker, erotic undertones of the machine's existence were driven into fiction, where they flourished. In Mary Wollstonecraft Shelley's prescient 1818 novel, *Frankenstein; or, The modern Prometheus,* the human-machine hybrid becomes a vehicle for preexisting cautionary fables of scientific excess and vaunting human ambition. As in other machine myths, the one we now call "Frankenstein" had earlier and elsewhere been associated with the devil's magic (as in the Faust legend) or cabalistic power (as in Jewish tales of the Golem). As the end of the nineteenth century drew closer, and the mechanical reproduction of labor and images became ever more extensive, a raft of authors emerged to explore the terrifying or erotic potential in the "hidden" world of machines. In the mid-1950s, the French scholar Michel Carrouges traced a structural continuity in these fictions, examining a group of what he identified (using Marcel Duchamp's phrase) as "bachelor machines." The patterns that emerge from Carrouges's analysis are useful in understanding the history of painting machines, for, perhaps not surprisingly, the painting machine tradition is strongly linked to these turn-of-the-century bachelor machines.

For Carrouges, the bachelor machine is any apparatus in which energy is closed in a circuit upon itself, and normal biological procreation is denied. Although many of the fictions he examines identify aspects of their devices as female (much as Duchamp's *Large Glass* also has a bride), Carrouges classifies them all as "bachelors." Within this structuralist analysis, Carrouges argues that the psychological dynamic of Edgar Allen Poe's tale of mechanized torture in "The Pit and the Pendulum" is continuous with Franz Kafka's inscriptional death machine in *The Penal Colony;* both evolve into the early-twentieth-century bachelors of Alfred Jarry, Raymond Roussel, and Marcel Duchamp (about which, more below). The structure shared by these bachelor machines displays an arrangement in which there is a space of discipline established above a space containing some disobedient element.[10] Some kind of man is suspended in the bottom space; he has broken the hierarchical codes of a regulatory society (the bachelor, the outlaw) and must await or endure his punishment from the disciplinary machine above (most clearly Poe's pendulum, but also Kafka's inscribing machine and, more tentatively, Duchamp's bride).[11]

Many of the women artists in this exhibition would disagree with the exclusively masculine bifurcation Carrouges envisions between the *bachelor machine* (which may paradoxically take the form of a mechanical "bride") and the disciplined male. Carrouges, writing in the 1950s before feminist critiques had made such views difficult to sustain, assimilates *all* these machines into the bachelor realm, simply because their forms of production and reproduction avoid the natural biological union of oocyte and gamete. In Carrouges's view, the realm of biology appears as somehow essentially female (despite the presence of male germ cells), and the world of machines is inherently male (brides though there be). His analysis, brilliant as it is, thus mirrors the views of the male artists and authors he discusses.[12] Although Carrouges first developed his analysis after World War II, his structure seems still to inhabit a late-nineteenth-century universe where men and women occupied entirely separate spheres of existence. Yet the very technologizing of sex represented by the phrase *bachelor machines* reflects the confusion and commingling of those separate spheres. The work in this exhibition shows the "maleness" of technology to be a social construction, one that becomes tenuous, at best, by the 1990s.

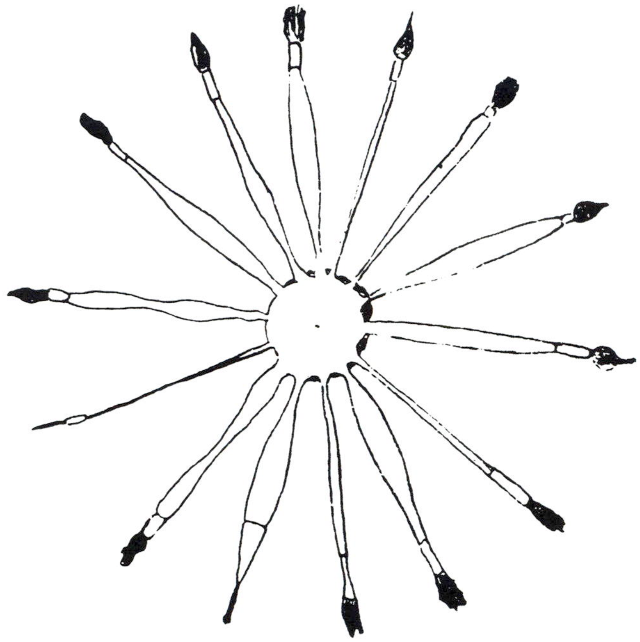

Even during the reign of men's and women's "separate spheres" at the turn of the century, there is evidence for more complexity in the gender constructed for technology. Duchamp's most famous work, *The Bride Stripped Bare by Her Bachelors, Even,* had contributed to Carrouges's concept of the bachelor apparatus. But at least in Duchamp's view, the machine has many feminine characteristics, and in *The Large Glass* (as it is also called) the bride has the upper hand.[13] Many artists besides Duchamp who came of age around the turn of the century drew parallels between a flourishing technological environment and the emergence of a bold new group of political and economic agents who happened to be female. A subliminal connection was formed between these newly active women and contemporary machines. Celebrated as New Women, trivialized as suffragettes, sexualized as flappers, or excoriated as *hommesses* and *garçonnes*, these active women took advantage of the shifting political and economic roles generated within industrializing societies to carve out new roles for themselves. By World War I, they were ready to take over important public duties in a society that had been robbed of its male population first by conscription, and then by death.

FIGURE 1.

Part of the painting machine described by Raymond Roussel in *Les impressions d'Afrique* (1910), as rendered by Jean Ferry in his homage to Roussel, *L'Afrique des impressions* (Paris: J. J. Pauvert, 1967).

Unsurprisingly, many men in a position to comment about this state of affairs were unhappy with the circumstances surrounding women's rise to power. Although everything from bicycles to dynamos was invested with imaginary female attributes, the control and mastery of those machines was reserved, whenever possible, for men. This is the sense in which the nonprocreative and disciplinary apparatuses that proliferate in the literature from the end of the nineteenth and beginning of the twentieth centuries can indeed be described as *bachelor* machines. For whatever their "sex" may be, they are always embedded in a system of social and economic relationships between men.[14] The rules of the family, and the supposedly biological imperatives of marriage (the domains of women and husbands), are ignored or flouted in these Duchampian cultural constructions—rendering them "bachelors" even when women are integrally connected with the apparatus at hand. And indeed, as noted above, many of the turn-of-the-century "bachelor machines" turn out to be part female in their operation—particularly those in which art plays a role in what is produced.

It is no surprise that these female operations occur most frequently in those fictions and artworks tinged with the imaginary erotic potential of the machine. We have already seen the fantastic worlds that existed as parallels to the rational domains of the Enlightenment; these worlds were few, and went largely unnoticed. It was only during the peak of the Industrial Revolution (the latter half of the nineteenth century, and first third of the twentieth), that such machine dreams exploded into literature, and later into visual art. It is worth exploring them in greater detail, for these fin de siècle entanglements represent the historical legacy animating much of the work in the present exhibition. Not all of these historical machines fit into Carrouges's structure. There are other narratives from the same period that embrace couplings between mechanical entities, or between humans and machines—love, rather than discipline, rules these encounters. Witness E. T. A. Hoffmann's "Sandman" (1817) and the ballet *Coppélia* (based on another Hoffmann story), with their themes of mechanical belles; or *Les chants de Maldoror* (1868-70) of Isidore-Lucien Ducasse (a.k.a. the Count of Lautréamont), with its famous romantic encounter between an umbrella and a sewing machine. Last but certainly not least, there is Villiers de L'Isle-Adam's astonishing novel *L'Eve futur* (1886), in which the American genius Thomas Edison invents a sultry, well-spoken android as a gift for an aristocratic friend, designed to replicate and replace the friend's beautiful but oh-so-vulgar real-life lover. These narratives emphasize substitutions of the controllable machine for the vexing female subject; not unexpectedly, creativity is reserved for the male inventor behind the scenes.

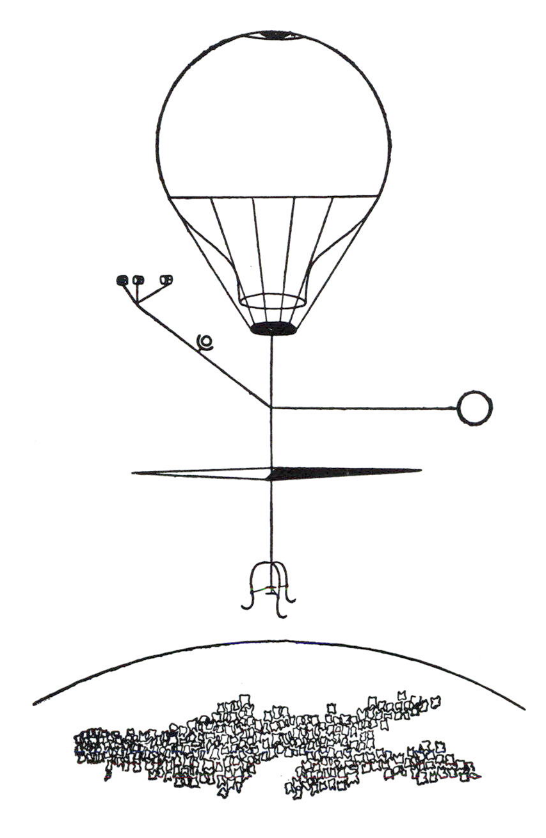

Figure 2.

Raymond Roussel's mosaic-making apparatus, *La hie ou la demoiselle,* from his *Locus Solus* (1914), as rendered by Roger Aujame in Michel Carrouges, *Les machines célibataires* (Paris: Arcanes, 1954).

Less central to their plots, but playing major dramatic roles that bear considerable interest for our theme of "painting machines," are the creative devices in Alfred Jarry's *Docteur Faustroll* (1898) and *Le surmâle*, "Supermale," (1902), and in Raymond Roussel's *Les Impressions d'Afrique* (1910) and *Locus Solus* (1914). Roussel's "impressions" were staged in Paris in May 1912, forever altering the imagination of the young Marcel Duchamp, who said he owed his *Large Glass* entirely to Roussel. The bizarre narrative by Roussel features a bed that generates photographic images, and a painting machine prepared by a woman, Louise Montalescot, whose lungs are pierced with surgical needles so that her every movement elicits a musical wheeze; this painting machine is visualized in Jean Ferry's homage to Roussel, *L'Afrique des impressions* (1967), where it appears as a paintbrush-studded wheel (fig. 1). The machine is to be manipulated by a spherical automaton daubing paint from a traditional palette onto an equally traditional canvas; its role, as programmed by Louise Montalescot, is to duplicate the beautiful trees of the surrounding landscape.

Roussel's second astonishing painting machine can be found in *Locus Solus* (either "lonely place" or "country of the sun"). Again, the machine has female components, although in this case it is controlled by the master of an estate where it floats as a magical amusement. It is female by virtue of its name, *La hie ou la demoiselle*, which in turn functions as part of a complex chain of puns—puns being the engine of Roussel's arcane method for generating pure fiction.[15] This art-making apparatus is a skeletal device suspended from an airborne balloon (fig. 2). It moves over a horizontal surface in the master's garden to make a ravishing mosaic that will depict a melancholic knight. Made entirely of human teeth of different natural colors (brown, yellow, bluish-white), the mosaic is set down according to the consummate program laid out in advance by the *demoiselle's* male master. Where Roussel's earlier painting machine separated the female producer from the entirely mechanical apparatus, this entirely female machine unites the two. Powered by wind and sun (whose patterns are known and mapped beforehand by the controlling master), this later *demoiselle* marks the moment of hybridity (both nature and apparatus; both female and machine) that appears a few years later in the poems and drawings of Francis Picabia, where the machine appears as *la fille née sans mère* (the daughter born without a mother).[16]

Less well-known (it certainly went unremarked by the isolated Roussel) was Jarry's painting machine, buried in the text of his *Gestes et opinions du Docteur Faustroll, pataphysicien*. This far more sexualized mechanism, despite its birth almost a century ago, appears closer in spirit to the present-day painting machines that serve as the focus of the current exhibition. Jarry's device churns through a deserted Hall of Machines, alone in the abandoned city of Paris, the only thing moving in the silence at the end of the world. Turning, tilting, and bumping crazily along like a top losing its momentum, it squirts "on to the steep canvas of the walls a succession of basic colours ranged according to the tubes in its belly, like in a *pousse-l'amour* in a bar . . . the unexpected beast, ejaculat[ing] to the walls of his universe."[17] The paintbrush-studded wheel of Roussel, the squirting mechanism of Jarry, the abandoned Hall of Machines at the end of the world—these are themes and images that seem familar within the imaginary universe inhabited by painting machines. As the works in this exhibition indicate, this chain of linkages from the nineteenth to the mid-twentieth to the late twentieth century holds up very well. It is only in the gender of the machines' primary programmers and producers that we see a dramatic change.

trajectories

Mechanolatry has an evolution, and it is neither seamless nor unidirectional. In the visual arts, there are three broad phases that can be identified: the *iconic* (portraits of machine forms), which extended from the sublime industrializing landscapes of the late nineteenth century through the first half of the twentieth; the *performative* (productions involving mechanical processes), which saw its peak in the decades immediately after World War II; and the *postmodern* (self-conscious manipulations of industrial images and processes), a vexing but inescapable term for the sophisticated, knowing modes of machinic art-making that began after the 1960s and continue to the present day.[18]

What this trajectory presumes, first of all, is that mechanolatry has never really disappeared. If at first it seemed that the machine had been demoted or demolished in the mid-twentieth century, upon closer inspection we find that it merely became more human ("passing," perhaps, and thereby less visible in our culture). Its "meat" became animated by unruly and creative impulses, and its apparatuses made gendered and generative in increasingly fantastical conjunctions. The desacralization of the machine—and its purported death at the end of the 1960s—was exaggerated. The rich subliminal and imaginary life of the machine went on.

True, the United States (precisely at the moment of that 1968 MOMA show "at the end of the mechanical age") was shifting away from being a primary industrial producer to becoming a secondary consumer of other countries' industrial goods. (The 1995 closing of the Bethlehem Steel Mill was but a delayed echo of this shift to a service economy.)[19] But the productive capacity of the machine is only one of its attributes, as this exhibition and essay make clear. As we have seen, even Descartes and Diderot may have had other things besides productivity on their minds. Going back well before those Enlightenment thinkers to Aristotle, the Roman Vitruvius, and Philo of Byzantium (to mention only a few examples), artists and thinkers have found in the mechanical an irresistible mirror, metaphor, and monstrous exaggeration of the human—as well as a potentially perfectible alternative to biological existence. Perhaps by the mid-1960s it seemed on the face of things that the machine had come to the end of its materially productive life, shipped off to the retirement home by that preening adolescent, the "electronic brain" (exemplified by the UNIVAC computer). But this was merely a forced removal from the front lines of the economy. The machine may have been the first victim of downsizing, but it remained embedded in our collective industrial imagination. By the early 1980s, it was clear that many artists (such as the ones seen here) had barely begun to tap the symbolic potential still dwelling in mechanical forms.

What was the effect of this shift? The Greek gods had passed into a golden age when they entered narrative mythology. In a similar fashion, although the "end of the machine age" seems to have bequeathed us a rich trove of meanings and images for visual manipulation, they often appear gilded by the honeyed light of a bygone era. The retrospective cast now given to many of the machine's meanings and images is different, then, from previous references to the machine in visual culture. The artist's machine may still appear futuristic on occasion, but just as often it seems to dwell in a lost age characterized either by its innocent optimism, or by its malevolent will-to-power.

Such a retrospective cast for this post-'60s imagery may be inevitable. We are, after all, nearing the end of both a century and a millennium; we have had a close relationship with the machine for much of that time. Intriguingly, as our relationship with machinery attenuates, it is becoming more visual (and visceral, if we take the interest in three-dimensional painting machines as analogues for the body—a logical outgrowth of the machine's demotion to the status of meat). While the nineteenth century witnessed the proliferation of literary investigations of the machine, the twentieth is characterized particularly by its visual and visceral infatuations with technology. Mechanomorphic paintings, drawings, prints, sculptures, and even architecture ebb and flow throughout the century. In its first phase, mechanolatry produced icons for its cult.

Such icons dominate the purely futuristic moment of the 1910s and 1920s, when the machine had moved out of the industrial sector to colonize the pedestrian realm of the street and the domestic sphere of the home. By turns fantastic, terrifying, and intoxicatingly powerful, the machine in this First Machine Age was seen to provide a new mode of organizing the human spirit. Henry Adams posed a famous choice for himself in 1906, between the Virgin and the Dynamo.[20] It seemed clear by the end of World War I that the Dynamo would be chosen (assuming, at times, some of the Virgin's female attributes). Yet for all their infatuation with the machine, visual artists of the early twentieth century approached it in largely iconic terms—as an awe-inspiring *image* first and foremost. While it was the mechanical systems of propagation and reproduction that engaged the period's writers, visual artists seem to have been struck primarily by the machine's powerful, dynamic, or glittering appearance. One thinks, for example, of the force lines of the Futurists, which echo and repeat a static fuselage moving through space (rather than show successive phases of a piston, for example), or the brilliantly centrifugal collages and assemblages of the Dadaists (which, when they use machine forms, treat them as semiotic signs rather than processes), or the gleaming metallic cylinders of Léger's Cubism (which brilliantly express a pure, hardened exteriority, described by Walter Benjamin as the "hoped-for metallization" of the human form).

Only artists of the Bauhaus, with their cool geometries and industrial linkages, or the Russian Constructivists, desperate to modernize humanity with their symbolic forms, would attempt to approximate the "Taylorized" and "Fordist" modes of production in their art.[21] These American concepts of industrial production had spread rapidly throughout Europe in the wake of World War I. They advocated that human movements be analyzed and broken down into repetitive, replaceable units, assembly processes routinized into regulated linework, and components standardized as interchangeable parts. Although postmodernism has given much of this drive toward productivity a negative cast, at the time it seemed positively utopian in its promise of human advance. The artists most radical in their attempts to approximate this rationalization of human production in their art were outside of what we construct as modernism's center. They worked far to the east of Paris, in Moscow, Dessau, or Berlin. Bauhaüsler Oskar Schlemmer, for example, organized a mechanized theater of repeated, geometricized, narrative-free movement; his colleague László Moholy-Nagy attempted to make a systematized geometric painting by dictating instructions over a telephone.[22] The Russians, still farther to the east, critiqued their old artistic culture and tried valiantly to achieve a new "Productivist" industrial art.[23] But dreams often outpaced achievement during this first phase of visual artists' infatuation with the machine, and those attempting to enter directly into industrial production in the early twentieth century appear to us now as naive.[24] In any event, most were content merely to portray industry from the outside; ambivalent or adoring, they gazed from afar.

The engines and generators in paintings from the first part of the century are rendered as exotic aliens, visitors from the future who appear the products of a miraculous birth. At the same time, these miraculous creatures are close to us. They are human-made, after all, and yet we remain surprised when their function or appearance betrays that they are literally produced "in man's image."[25] The machine's mirrored surfaces reflected a divine potency for the early modernists, but the Divinity was us. Even the Surrealists, who aimed at questioning this complacency more thoroughly than most (and became the forces behind the revival of Roussel, Jarry, and de L'Isle-Adam), remained "outside" the machine in their art, the better to gaze at its flattering reflections. Their carefully, even academically painted canvases apply traces and signs of libido to the mechanical form as if they were adding a superfluous flywheel to a perfectly functioning machine. Few probed beneath this metallic skin to explore the more complicated systems and internal workings of the mechanical life within.[26]

The second performative phase of mechanolatry's evolution occurred after World War II. It was a radically different context, in which the machine was both ubiquitous and invisible, part of a thriving industrial order that dissolved itself into the "corporate identity programs" saturating visual culture. The new, pervasive corporate logotypes adopted the iconicity of earlier mechanomorphic art, reducing their narrative messages ("Bell Telephone" and "Chase Manhattan Bank") to gleaming, abstracted, wordless geometric forms.[27] The ubiquity of such corporate icons found its echo in the art: the metallic-colored die-cut shapes of Stella's Aluminum Paintings, or the crisply silk-screened graphics of Pop art. In addition to this redoubled iconicity in both art and the corporate sector, however, a new type of industrial aesthetic became common among artists, a mode in which the production of art was explicitly analogized with the production of common manufactured goods. The artist would now *perform* as mechanically as possible (besides producing icons of machine-made forms). This new love affair with technology took place at the level of systems and productive forces, getting under the machine's metallic skin to get, as it were, to the real "meat" of it all.

In North America, and increasingly in Europe and industrialized Asia as well, the dominant backdrop for this "performative" postwar art production was a gestural style of painting that emphasized the existential position of its male creator. The central example of this way of being was the American Abstract Expressionist painter Jackson Pollock. His painting method, itself a kind of performance that could never consciously be experienced in those terms, was interpreted as an unmediated encounter with the prone canvas. Photographs and films showed that no preparatory drawings or sketches were made: Pollock moved around the blank canvas, dipping sticks into cans of enamel and flinging paint on the unprimed duck. Pollock's performance (and for him it was resolutely unselfconscious) encouraged painters to think of themselves as being artists through their *actions* rather than their objects. The "performative" was born, and it would not be long before it was allied with the mechanical, at least in the United States.

Besides American Abstract Expressionism, many other modes of gestural painting emerged around the globe after World War II (forming the generational "fathers" of the artists in the current exhibition). In Europe there was Tachisme ("blot" or "stain" painting) and Art Informel, and later the CoBrA group. In Japan there were the actions, paintings, and sculptures of the Gutai Group. Though often explicitly "performative," these modes of painting were resolutely antitechnological (with the exception of certain Gutai artists, as we shall see). The artists of the late '40s intended their works primarily as assertions of humanity that confronted the all-too-technological and antihumanist machines of war. As the immediate shock and outrage over those war technologies waned, the United States' military infrastructure became the basis for the cold war explosion in scientific and technological research, evolving seamlessly from the top-secret Manhattan Project into the massively public drive to be the first nation to land men on the moon. In this revised, more technologically optimistic context of the late '50s and early '60s, the angst-ridden, individualist gestures of the New York School painters (and their cohorts across the two oceans) began to seem strained and false. The eclipse of mechanomorphic icons had begun, and when machine dreams reemerged, they did so with a uniquely performative emphasis.

Tinguely, whose self-destroying kinetic sculpture we encountered earlier in this essay, was one of the first to critique the assumption of mastery that the gesture was held to proclaim, and to tie "performance" to the machine. In a series of sardonic drawing machines of the late 1950s (dubbed "Meta-matics" by curator Pontus Hultén), Tinguely intimated that the supercharged, authorizing touch of the artist's brush (linked to handwriting through the concept of the "signature" style) could be duplicated by a simple assemblage of a motor and a marking device. These extraordinary kinetic hybrids were small, tabletop, metal sculptures, welded together and decorated with whimsical, Alexander Calder-like fillips and curlicues; they were exhibited most memorably in a large conglomeration at the Iris Clert Gallery in Paris in 1959, and again at New York's Staempfli Gallery in 1960. But unlike Calder's gently air-driven mobiles, each of Tinguely's Meta-matics bears an obviously visible motor (sometimes triggered by a coin) that sets it churning into action (fig. 3). The ambiguity begins with these gyrations, for the machine is not content with motion alone (a "dancing sculpture"), but seeks to produce drawings by jabbing markers at a page. Despite the Meta-matics' powerful critique of creativity, however, human collaborators appear to be crucial—selecting the pen, positioning it correctly, ensuring that the paper is clipped firmly to its stand, and activating the motor. Human collaboration was further suggested by two other countervailing facts: the drawings were often signed and removed by the collaborators as "made by them," and Tinguely destroyed most of the Meta-matics "not because they are ugly but because they draw badly."[28]

In almost eerie synchronicity with Tinguely's Meta-matics, but totally independently, Japanese Gutai Group member Akira Kanayama produced a number of automatic and remote-controlled painting devices in 1957. In Kanayama's pieces, tiny robots crawled across a prone canvas and dribbled paint in mechanical mockery of the Abstract Expressionists' expansive, angst-ridden gestures. The Gutai Group had formed only a year earlier, with the aim of merging certain aspects of traditional Japanese art (such as calligraphy) with a more contemporary spirit (like that which Abstract Expressionism seemed to represent). Where the Abstract Expressionist gesture had encoded a "signature style" marking an individual personality, the calligraphic line in Asian tradition represents nearly opposite characteristics, embedded in religious and aristocratic disciplines of self-effacement. Thus, as well as mocking the apparent egotism of Abstract Expressionism (and the Tachist gestures of French painter Georges Mathieu, who had been to Japan that same year), Kanayama's robots also returned, if somewhat ironically, to Japanese Zen Buddhism—the machine's patient movements serving literally as a mechanism for quieting the self. Add to this the postwar explosion in Japanese industrial technology (these were the years when every mechanical windup toy was "made in Japan"), and the grinding demands of the Japanese workplace, and Kanayama's robot calligraphers become conflicted investigations of the contemporary Japanese way of life.

A few years later, on the other side of the planet, evidence of another painting machine appeared. The Situationist Giuseppe Pinot-Gallizio installed in a European gallery a seemingly endless roll of paint-spattered canvas, titled *Industrial Painting*. Like the products of Kanayama's robots, what one could see of Pinot-Gallizio's painting also looked like the surface of Jackson Pollock's skeins. But Pinot-Gallizio's canvas was only partially unrolled from a sturdy metal rack, spilling onto the floor and implying that it went on forever. As in Kanayama's piece, it bore a complex relationship to the impoverished Italian economy—as if "mass production" consisted of this single bolt of spattered cloth.

In tandem with these productions made by men (but far less visible), a tradition of woman-authored painting machines began to emerge. One might point here to the work, in the late 1950s and early 1960s, by Niki de Saint Phalle and Carolee Schneeman, both of whom interrogated in interesting ways the interface between the artist's body and the technological. In *Shoot Paintings* Saint Phalle—a French-American artist who was involved with Tinguely at the time—used rifles to "automate" the process of coloring her plaster paintings in a series of dramatic performances that began in 1962. Underneath the bumpy but otherwise dead white surfaces of her plaster panels were suspended vessels of colored paint: when these were shot at successfully by her collaborators, color exploded from within the plaster, creating a vibrant "gesture" on the picture's surface. As well as aping the appearance of Abstract Expressionist and Tachiste canvases, Saint Phalle's *Shoot Paintings* also dramatized the implied violence of "Action

Painting," as Abstract Expressionism was also called. Carolee Schneeman, active in the same group of young cosmopolitan artists, made painting machines closer to those in the current exhibition. Influenced partly by the occasionally electrified assemblages of Robert Rauschenberg, as well as the moving parts in Joseph Cornell's boxes, Schneeman made kinetic assemblage paintings in which the moving parts evoked art-making gestures. This interest continued through the mid-'80s, when she rigged a mop to a mechanism that caused it to slap the floor at dramatic intervals—a sporadic, mechanical, "Action Painter."

The look of the art and processes in these various endeavors initially bore an ironic resemblance to the dominant style of Abstract Expressionism. The gestural abstractions that came out of Tinguely's Meta-matics and Saint Phalle's *Shoot Paintings* seemed superficially to be as passion-ridden as the countless canvases then being produced by followers of the New York School (although Tinguely's more closely resembled the energetic drawings of the European Tachist Hans Hartung, and Saint Phalle's looked like more spontaneous versions of a canvas by Georges Mathieu). The "automaticity" of these mechanized painting processes presented an ironic commentary on everything from Enlightenment automata (in particular, the Swiss "Draughtsman" from the eighteenth century) to Surrealism's much different "automatism," that trance-like state intended to release uncultivated imagery from the depths of the subconscious. One could not help sensing in such mechanized art-making that the performative body of the artist could easily be replaced by the machine. The artist's creative body was no longer divine. Like the machine, it had become just so much meat. The art lay elsewhere, in the union of concept, performance, and the object itself.[29]

Figure 3.

Jean Tinguely, Meta-matic installed in the Iris Clert Gallery in Paris in 1959. Photograph courtesy Pierre Matisse Gallery, New York.

From this union came the third phase of art that this exhibition documents, where iconic and performative mechanolatry appear at the same time, but within a context of art production that is determinedly postmodern. In the current frame, the matrix of references and meanings are more complex and multilayered than they were in the 1960s. Rather than acting "like" a machine, as some from the early '60s had sought to do (Andy Warhol chief among them), the postmodern artists seen here present the machine either as literal stand-in for the body, or as symbol of the late-twentieth-century urban environment. In this sense we have entered a "post-performative" phase, where the artist's intention is neither as accessible (there are no Warhols declaiming, "I want to be a machine"), nor as crucial to interpretation. By the same token, the painted image no longer remains within a local frame of reference, but ranges through the vast domain of reproductions that blanket reality in our present-day world.

The reemergence of mechanomorphic form and process during the past fifteen years is thus neither a revival of early-twentieth-century "Machine Age" art, nor a revision of the mid-century involvement with the postwar economic order. The German and American artists in the current exhibition belong to a generation positioned at the end of "the American century" (as Henry Luce dubbed it during World War II), working in an economy identified as both postindustrial and postcolonial, where capitalism reigns globally and electronic webs circle the earth.[30] The new configurations of the technological "cultural imaginary" that appear here reveal the role that mechanistic models of production play in what one might call the "unconscious" of painting—the fundamental desires that motivate aesthetic need. These artists cannot access the machine with the naive optimism of the Soviet revolutionary, nor can they muster the pure will-to-power of the early capitalist entrepreneur. Their works must use machines to speak about specific historical moments and historically bounded activities—looking back at *modernist* moments and activities in which seemingly neutral ideas of progress concealed deeper needs for technological power and mastery. As the sophisticated artists in *Painting Machines* demonstrate, such dreams of mastery are intimately tied to the art-making impulse itself, just as the desire to make art is linked, deep in our psyches, with a love of technology as a way of making and unmaking the world. Exploring our emotional, aesthetic, and symbolic investments in machines, the artists in this exhibition make works that resonate with the largest issues of technology's place in contemporary culture.

painting MACHINES

In selecting work for the exhibition, we decided to emphasize those artists who used the machine as either the functioning substance or meditative subject of their art. In both cases, priority was given to uses of the machine that commented on larger issues concerning the place of technology in culture, rather than machines that served as vehicles for other messages having nothing to do with technology or art per se.

Art that was merely "kinetic" (albeit machine-driven) was not enough: for this exhibition, works had to comment on the art-making impulse itself. Machine-makers in video, computer, or other electronic media—even if their works commented about art-making, technology, and culture—were also excluded. Photography was disallowed for the simple reason that its investigations of industry have already been amply documented, most recently in the Boston University Art Gallery's own exhibition *From Icon to Irony*. These binding decisions about medium and genre came also from the view that these newer media are themselves our most "technological" art forms, and thus their discursive relationship to industry is even more complex than the "simple" machine-age media we chose to investigate instead.[31]

Following criteria similar to those governing the choice of sculptors and other "technologists," painters of machines were burdened with the task of making some comment on the industrial subjects they chose to depict, and on art's conflicted relationship to those subjects. There was no presumption that the comment had to be either negative or positive, just that the paintings concerned went beyond a naturalistic rendering of machine forms to raise issues about those forms (and forces) in our lives.

Within these parameters, the working selection for the exhibition revealed some interesting demographics early on. The majority of the artists (Gipe, Larner, Moskowitz, Sultan, Tansey) are American, but two are German (Horn and Trockel) and the youngest (Bulloch) is Canadian by birth and now lives in London. All of them save Moskowitz are "baby boomers," part of that population surge beginning at the end of World War II and extending its bell curve through the 1960s (Larner, Gipe, and Bulloch standing at the very outside of that curve with birthdays in 1960, 1962, and 1966, respectively). Most of these artists grew up with the peculiar relationship to technology that was characteristic of life during the cold war: technology would blow us all up, or send us to the moon; it caused cancer, or offered the only hope for cancer's cure; it was the source of all meaningful work, or exactly that which would render traditional skills obsolete.

The resulting group is something of a cohort, and the mid-1980s witnessed their greatest production of mechanomorphic art (whether they were just beginning to produce, or found themselves in successful mid-career). Most significantly for as gendered a realm as the technological, all the makers of machines turned out to be women, all the painters of machines, men.

Of course, all of these interesting data are in part the artifacts of a minute sample, further torqued by the limitations of space and available loans. Men *do* make machines, and some of them are painting machines; similarly, women do paint, and sometimes they paint machines.[31] The gender boundaries this exhibition documents seem to have been more rigid in the 1980s, however. It is only recently that one can name a greater number of cross-gender examples of the trend identified here. In 1994, for example, British sculptor Damien Hirst presented an installation at Berlin's Bruno Brunnet Gallery in which visitors were invited to join him in "Making Beautiful Drawings" by activating a fairground "spin art" machine.[32] And in a 1995 exhibition by Canadian painter Medrie MacPhee, depictions of industrial fragments were seen to comment on "the ruin or collapse of modernism," a concept familiar from the elegiac views of male painters shown here.[34]

With equal certainty, there are artists in places other than Germany and North America who pursue both modes of engagement with the image of technology. But perhaps it is not surprising that broad trends in recent art, and some of its best-known practitioners, fall into the patterns suggested by this exhibition: that the two major postindustrial giants (Germany and the United States) should produce artists interested in earlier phases of mechanolatry in their national pasts, and that gender should have a strong influence on one's relationship with both art and technology. First, let us examine this national divide.

The strength of Germany's "iconic" phase of machine-age art is obvious from even the most cursory discussion of the Bauhaus and its enduring dreams; the solitary magnificence of Konrad Klapheck's chilling, funny tableaux from the same period remain too little known. His paintings bear loaded titles, such as *The Perfect Husband* (a typewriter shown in looming close-up), *Intriguing Woman* (1964, a sleek industrial sewing machine), and *The War* (1965, ranks of forbidding machines that look not like panzers but like bandsaws). Klapheck's style looks familiar, as a kind of updated *Neue Sachlichkeit* (the "New Objectivity" of the 1920s), where the machine icons take on a sinister air. But less well known on this side of the Atlantic is the "performative" second phase of Germany's industrial aesthetic, which appeared in the 1960s, at the same time it was emerging in the United States. Many are now aware of the important work of Sigmar Polke and Gerhard Richter, but few remember their "Capitalist Realism" in the early 1960s. This more biting and sardonic variant of American Pop art first appeared in a Düsseldorf department store, when both artists posed themselves and their paintings as part of the industrial art and furniture on display.

Combining these postwar German exemplars of industrial critique with Joseph Beuys's overwhelming artistic and conceptual freedom, Rosemarie Trockel and Rebecca Horn have a rich tradition to which they can allude and every reason to make such allusions in sculpture. For Beuys, every person could be an artist, and every social interaction could become art. While he preferred the archaic persona of the shaman in his self-presentations, Beuys was equally willing to muster industrial crane crews and hydraulic pumps in the pursuit of his aesthetic and political projects. Interestingly enough, the German-born artist Hans Haacke, although he pursued his artistic career entirely in the United States, nurtured views very much like Beuys's, in which technological and natural systems were aesthetically useful only if they could be manipulated by the artist to pose larger questions about society. Liz Larner and Angela Bulloch, although more proximate to North American influences like Haacke, have clearly learned from this German tradition, probably through the mechanical sculptures produced by Trockel and Horn themselves. Unlike the more geographically isolated generation of Haacke and Beuys, these younger artists have come of age in an artworld that no longer divides itself so unproductively along national lines. Larner could see Trockel's retrospective in California in the mid-'80s, and Bulloch could meet Horn during the years when both were living, working, and exhibiting in New York. In the period this exhibition chronicles, the "market wall" between Europe and the United States had begun to crumble, not unlike the Berlin Wall at the decade's end.

If the tender young tradition of painting machines seems linked primarily to Germany (despite extraordinary French precedents), perhaps it is equally predictable that many of the Americans in this exhibition should have made a commitment to the painted rendition of mechanical form. With a briefer aesthetic past to draw upon, American memories understandably turn to the accomplishments of Precisionist painters of the 1920s and 1930s (artists such as Charles Sheeler, Ralston Crawford, Georgia O'Keeffe and Charles Demuth); indeed, Sheeler is an explicit model for at least one of the artists seen here (Lawrence Gipe). Together with the complex international influence of New York Dada (Francis Picabia, Man Ray, Marcel Duchamp, and others), these are powerful legacies for American painters who are drawn to machines and their environments as subjects for art.

Another powerful national tradition is offered by the discourse of the Sublime, which became, in the United States, something of a national religion. This spiritual experience begins with awe, moves through terror, and culminates in the mastery of these emotions; supposedly, it is not *depicted* by the artwork but *engendered* by it. Where nineteenth-century painters could construct an awesome wilderness to build their visions of America, with only the merest hints of railroads, lumber trails, or homesteads as tokens of the March of Progress, these late-twentieth-century artists see nationhood retrospectively, as an equally awesome industrialization of that same "virgin earth." That they complicate and question such achievements comes with their self-consciousness as *post*modern artists. Gipe, Sultan, Tansey, and even the older painter Moskowitz all place their industrial scenes within imaginary quotation marks. Theirs is not a photographic realism, although their paintings may mimic photography's laconic tone. Similarly, although relationships of size and scale in their works openly court Sublimity, they are careful to pose that rich Romantic tradition against a cooler ironic tone, achieved variously through the devices of monochrome, titles (Sultan and Tansey), interior captioning (Gipe, see page 22), inexpressive paint-handling (Moskowitz), or disjunctive representational effects (Tansey).

These painters of the machine further bracket their views of technology in the themes they choose, acknowledging technology's seductions even as they attempt to articulate the problems intrinsic to its use. Gipe shows networks of complicity and greed that link industrialized nations around a complex, politically divided globe. Sultan explores the principles of chaos and entropy that lurk in every organized system of energy (of which industrial technology is among the most complex). Tansey demonstrates how machines not only obey the obvious laws of national self-interest and solid-state physics, but also follow the patterns of our productive brains. Moskowitz, too, holds his imagery at arm's length, interrogating the industrial icon as a phallic signifier of modernity's abstract, monolithic power. As noted earlier, many of these painters present their visions in elegiac tones. For all its upright potency, Moskowitz's empty *Stack* speaks also of stilled factories, its palette a close-valued nocturne of melancholic hues. Similarly fuscous in their colors, Gipe's lush scenes glimmer from crepuscular shadows, and Sultan's paintings of disasters display torched and tortured surfaces that flicker with a sulphurous light. These are paintings that mourn the loss of productivity, rather than mimic its plenitude.

Despite (or perhaps because of) their mordant humor, the painting machines in this exhibition offer far sunnier commentaries on technology's future. These German sculptors (and the two Anglophones) seem to inhabit a more optimistic, or at least a more active, universe. What we can laugh at, we can control (at least conceptually). The tragic, by contrast, speaks only of those histories that lie eternally beyond our abilities to amend or repair them.

More tantalizing than this national gap between the mostly European makers of machines, and the largely American painters of Sublimity, is their gender divide. With few exceptions, those who make painting machines these days are women; those who paint our decaying industrial sites tend to be men. Put most simply, the women who construct machines that ape artistic operations place the machine in the position of the artist's body, specifically in order to critique the historical formula by which (male) gesture equals (male) genius. Just as simply, the men who paint pictures of machine forms seem to be questioning more general relationships between society and technology (or men and their machines), at a time when many men are experiencing a loss of control over the mechanical universe.

The robust quality of this gender divide undoubtedly results from the gendering of technology in the larger culture. Interestingly, machines are among the few categories of inorganic objects given gendered pronouns in English (as in "my ship . . . she"; "my car . . . she"). The implied operator of a putatively female machine, however, is always male. Cars, trucks, tanks, motorcycles, and airplanes have long been understood as tools for cementing relations among men (accounting for the strange irrelevancy of the female love interest in the film *Top Gun,* and the famous superstition against having women on board a battleship). While they may be referred to, in a proprietary way, as "she," such consumer goods as cars and motorcycles are advertised as phallic fetishes, the obligatory sultry female posed winningly against their windshields. With the subtlety available only to advertising executives looking for an unfilled marketing niche, current industry wisdom has it that games for the computer are so pervasively masculinized because "for him it's a toy, for her it's a tool."[35] Designed to flatter—pointing up the supposedly greater studiousness of little girls—this statement unwittingly parallels the syntax of mastery that industry itself proclaims: men will be technology's masters, even if women will constitute the work force that puts it to use.

Luckily, to the artists here technology presents a far more subtle face. Both the paintings and sculptures in this exhibition complicate the current gendering of technology in culture, and press hard on our attitudes about "who's responsible" for our relationships with machines. Through their media alone, the painting machines and machine paintings in this exhibition work on our sensibilities in ways quite different from the stereotypes offered by commercial journalism and advertising. Technology is neither "bad guy" nor "sexy commodity," but a deep and telling expression of ourselves.

Lawrence Gipe

Painting #4 for *Themes for a fin-de-siècle (Pride)*, 1989

Oil on wood panel

90 x 90 in.

Private Collection

As suggested in the brief history above, the iconic tradition (paintings of machines) is not necessarily older, but it is certainly more extensive than the tradition of machines that paint. The paintings here that explore mechanomorphic imagery begin with the reticent mastery of Robert Moskowitz, whose factory smokestack icons stand alone and unique in his oeuvre. Certainly they relate to his city skylines, but distilled and repeated until they have a portentious purity all their own. Stark and empty, *Stack* seems to loom like some skeletal fragment—a fossil from an age where ambivalence about "satanic mills" was always resolved in favor of building another one.

Surely *Stack* stands somewhere behind the dramatic industrial silhouettes of Donald Sultan. But Sultan situates his views in a messy, entropic reality. Not yet fossilized, his stacks may belch flame and smoke, or at least swirl in a cloud of fumes that are undoubtedly as noxious in life as they are beautiful in his representations. Bearing sumptuous, encrusted surfaces that engage the senses of touch and smell, as well as the more rarefied one of sight, Sultan's works employ some of the very materials of industry (roofing tar and asbestos vinyl tiles) that they depict. He constructs tenebrous, ambiguous scenes with titles that sometimes confess their relation to industrial disasters (as in his *Poison Nocturne*), and sometimes hide their status as depictions of forbidding way stations on a technologized journey toward death (*Polish Landscape*). Whatever we may glean from such a title as *Plant,* however, we intuit the work's brooding darkness, and savor the ironic contrast between the title's organic metaphor of growth and the painting's austere desolation. Sultan stands closer to his fuliginous landscapes than the discourse of Sublimity traditionally allows. (The humorous summary of Edmund Burke's magisterial treatise on the Sublime is: when you're on the mountaintop watching the flood, it's Sublime; when you're in the flood, you're drowning.) The level-headedness with which Sultan limns his scenes also limits and contains their sublimity, suggesting that post-'60s coolness so characteristic of the postmodern generations. This sense of distance constitutes these paintings' postmodernity: they are but intermediate stops along a signifying chain, not substitutes for reality.

Los Angeles painter Lawrence Gipe also aims for a cool, postmodern plateau in what Guy Debord called our Society of the Spectacle.[36] Gipe's landscapes of heavy industry follow a long tradition of American painters of the technological sublime, particularly Sheeler, whose smooth paintings of train engines Gipe has rendered in buttery impasto. Like Sultan, Gipe courts many of Sublimity's effects, but seeks to cast them in a postmodern, elegiac light. Across his sumptuous, thickly layered canvases, Gipe letters cautionary messages—"Power," "Desire," "Pride"—warning us away from the very seduction his images proffer. In his triptych focusing on the German Krupp family ironworks, Gipe plays this dangerous dialectic to the hilt. The painting's sumptuous brushstrokes and enormous size, the triptych format, and above all the turbulent and inspirational lighting all invoke religious altarpieces and spiritual landscapes. Gipe attempts to rein in these overwhelming pictorial effects with carefully ironizing texts above the images, German phrases and English words that hint at networks of complicity and guilt. Guilt emerged from "pride," which was a sin before it was a badge of self-esteem; and complicity is located in the relations between industrialized nations, where yesterday's war criminals (e.g. Krupp) were recycled into capitalist guardians against a Soviet victory in the cold war.

It is noteworthy that both Gipe and Sultan explore the Holocaust in their work, and even Moskowitz's *Stack* can be interpreted in its shadow. As we shall see, there are unavoidable echoes in some of the painting machines as well. Together with nuclear warfare, the Nazi death camps stand as the twentieth century's most horrific example of industrial-scale death. Rather than excoriate some monolithic structure of "Enlightenment rationalism" or "Western society" for these demonic technological developments, Gipe and Sultan seek to specify the place and the relationships that made such horror historically possible—even as they seek a wider and more general significance for their work. Sultan's train station is a Polish landscape of death, and Gipe's narrative of complicity is a complex tale of specific compromises and workaday business deals conducted monstrously, with human lives in the balance.

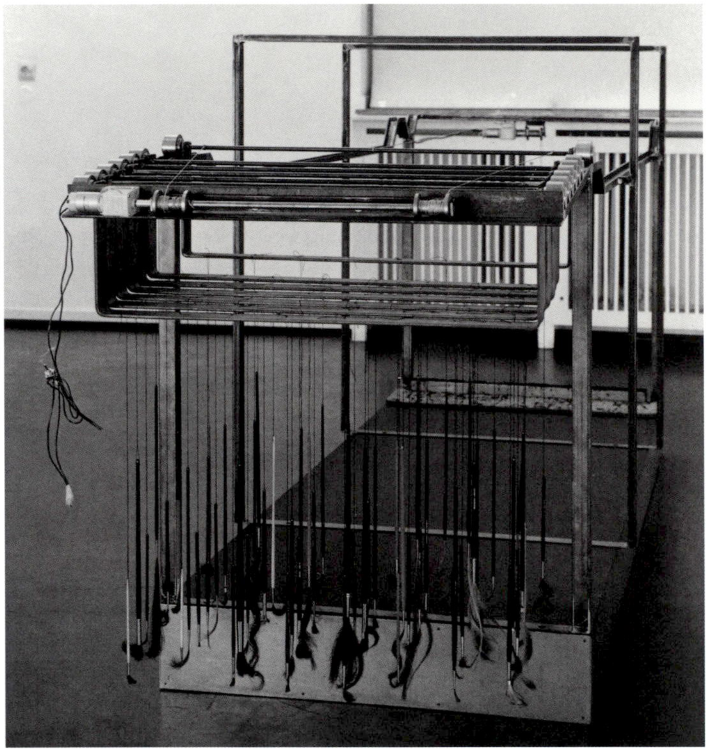

Unlike these painters, and more in tune with the women machine-makers, Mark Tansey takes a lighter, more humorous tone in his work. Like Trockel and company, he uses machines as metaphors and analogues of artistic production. The issue for Tansey is not so much the social position of technology, but rather its imaginary role as a figure for the artist. Recent works such as *The Matrix, The Raw and the Framed,* and *The Bricoleur's Daughter* visualize the (sometimes absent) artist as a consummate manipulator of technologies. This makes sense, as Tansey is himself a virtuoso technician. He has perfected a fiercely mimetic adaptation of Max Ernst's *frottage* technique, in which patterns are "painted" by subtraction rather than addition. A textured object or tool is pressed into wet paint on a highly gessoed canvas; when the object is pulled away, it removes the wet paint to leave a light trace behind it.[37] In his "toner drawings," Tansey intervenes directly in the internal workings of his photocopier to produce haunting montages of manipulated photographic images. The results are uncanny, appearing as some kind of photograph for which we are no longer sure there can be a source in the real world.

Figure 4.

Rosemarie Trockel, *Malmaschine* (Painting Machine), 1990. Steel, press plate, felt, and brushes from the hair of fifty-six artists. 101 1/8 x 30 x 45 1/4 in. Courtesy Monika Sprüth Galerie, Cologne, Germany.

The mechanical meditations by Tansey in the current exhibition join earlier, more sardonic paintings such as *Homage to Frank Lloyd Wright* (in which the Guggenheim Museum is constructed as a Blakean satanic mill). Bringing his critique of the "art machine" back home, *The Raw and the Framed* and also *The Matrix* suggest that the making of art is an everyday process where you show up for work, turn on the conveyor belts, and just try to keep up. Key to the phantasmagoric quality of Tansey's work, however, is his conception of the nodal site in which raw materials become art. In both *The Matrix* and the toner-drawing study for the large-scale painting *The Raw and the Framed,* semirefined matter (paint tubes, rock chunks) passes into a magical area, understood conceptually as the engineer's unexamined "black box," from which art emerges. The humor, of course, resides in the fact that by showing us "everything," Tansey has revealed nothing. His paintings *of* machines recall, in their scale and ambition, "les machines" (large history paintings *as* machines) of the nineteenth-century French salon. Equally mysterious in their gestation, they both rely for their conception on a complex process of image gathering, material manipulation, and "framing." Like the exhibition's *painting* machines, Tansey's postmodern history paintings present themselves as witty commentaries on their own coming-into-being—physical and yet fully magical.

Reversing and reflecting these pictures of moving machinery, the women artists shown here construct technologically fluent apparatuses. Their machines both present themselves as art, and work to duplicate processes of artistic production. These witty, provocative assemblages provide their own postmodern critique, directed this time not at society's reliance on technology but at artistic traditions of "genius" that both proscribe the use of technology and seek its totalizing effects. Rebecca Horn and Rosemarie Trockel dominate the recent history of painting machines, referring to earlier neoDadaist precedents (such as Jean Tinguely's Meta-matics) while critiquing "originality" and "mastery" in a thoroughly amusing way.

Trockel's *Malmaschine,* (Painting Machine—fig. 4), unfortunately unavailable for loan, is the magnum opus of the genre. For its dangling brushes, Trockel requested hair from such artist-heroes as Joseph Beuys and Cindy Sherman; each swatch of hair is bound neatly in a ferrule and attached to a wooden handle imprinted in gold with the artist's name. These exquisite brushes were dipped in inky black tusche, and moved by the machine down a length of fine paper; the finished works (*56 Brush Strokes,* included in this exhibition) were framed and labeled with the names of the successful artists—registering either that their hair was responsible for the image, or that the resulting lines constitute a "portrait" of the artists named. Potentially, the use of hair can be as disturbing in its associations as Beuys's use of fat—reminding us of the monstrous "harvesting" of such materials from victims of the Nazis' abattoirs. In Trockel's works, however, these associations are kept to a minimum, for the overriding discourse is that of painting. After all, these are not piles of artists' hair destined for wigs; they are exquisite tufts, closer in spirit to the locket-prize memento of a loved one (which has its own nostalgic discourse with death). Adding to the rich layers of meaning in *Malmaschine* and in *56 Brushstrokes* is their critique of the largely male rhetoric of painterly gesture. As noted above, many other modes of gestural painting appeared alongside the dominant postwar American discourse of Abstract Expressionism. Among these was a vivid German tradition—first in the heroic phase of German Expressionism, and then in its revival during the '80s by solemn German Neo-Expressionists like Georg Baselitz. Trockel takes on this tradition of the painterly gesture, first by automating it, and then by perversely personalizing it. In turning "signature style" from an expression of the embodied male individual to a product of what is literally part of the body, she forces some amusing conclusions. For the standard trope of the woman is her fulgent hair, a stereotype suggesting invidious comparisons between Cindy Sherman's thick "empowered" brushstrokes and the more tentative ones attributed to the balding Joseph Beuys.

Trockel continues her complex interrogation of gender roles and stereotypes in her knitted paintings and related apparatuses. In *Untitled (Woolmark),* she investigates the status of weaving as "women's work," posing this traditional view against the manifestly machine-knitted "painting" that displays the corporate woolmark as its only image. Just as machines have replaced the hearthside spinning wheel, the woolmark replaces the local patterns and designs with which women formerly signaled the unity of their folk. At the same time, Trockel's complex object asserts women's traditions of textile arts as *art,* stretching the knitted fabric as a painting, and positioning it in the place reserved for male genius. Although she has obviously mastered the technology of the knitting machine, in *Ich kannte mich nicht aus* (I am Stumped or I Don't Know About Myself) she offers a shrug of incomprehension: what she refuses to comprehend seems to be the infusion of mystical beliefs into the hapless machine. As Isabelle Sobin reveals in her essay for this catalogue (pages 53–54 below), the object shown in the photograph that hangs from the apparatus is a nineteenth-century device that was intended to prove, scientifically and mechanically, the existence of God.

Rebecca Horn's painting machines are less personal in their critiques of technological fancy and art-world hubris. They seem instead to echo the Luddite whimsy of the eating machine in Charlie Chaplin's *Modern Times* (Horn is a connoisseur of silent films). In her work, mechanically animated brushes gently stroke the air, and pigment cups madly dash their colors against the wall or scatter paint through open stretchers onto the floor. Such works constitute a mordant critique of the "possessed (male) genius" seen to undergird successful art from nineteenth-century Romanticism through to the Neo-Expressionists of contemporary art. When the possessed genius is a machine, of course, we can interpret it as merely dysfunctional. (Will somebody *fix* this thing?)

While they poke fun at notions of technological omnipotence, both Trockel and Horn present a more optimistic view of technology than do their male counterparts in this exhibition. In their work, it is clear that machines free artists—women artists in particular—from the baggage of the old "masterful" pictorial traditions. Far from bearing heavy existential weight, "gesture" in Trockel's *Malmaschine* is merely a seismic shudder among the slender suspended brushes; spilling through the open stretchers in Horn's *Kleine Malschule* (Little Painting School), gesture results in something of a mess. These painting machines offer an evocative, even surreal poetry. Like Diderot's crystal couplings or Descartes's animated belle, they have something other than economic production in mind.

The minimalist apparatus of Liz Larner's *Wall Scratcher* is equally quirky. Like Horn's early, pigment-free circle-inscribers, *Wall Scratcher* mechanically repeats the most primitive marking gesture. Gentler than Larner's contemporaneous *Corner Basher,* the *Scratcher* cuts into the wall with simple lines and indentations. Its patient persistence underscores the obsessive, cathartic, and potentially violent nature of such activity, and we learn in Ana de Azcárate's essay (pages 41–42 below) that the *Scratcher's* modest incisions have been compared to a prisoner's calendrical ticks on a cell wall. But the *Scratcher* is irrational—its marks are more like compulsive scribbling than any purposive attempt to "mark time," for the machine applies itself to the same spot for however long its curator determines. Like Angela Bulloch's drawing machine, Larner's is dependent upon human agents for direction. Both artists focus on the usually tacit power relationships between people and machines, and insistently bring them to the surface.

Larner has continued to pursue her interest in both human agency and mechanical metaphors. She is currently engaged in a complex project, the plans for which have been conveyed as a suite of drawn and printed collages called simply *Machine.* The final project will require extensive collaboration with sophisticated industrial machinists and computer programmers; the texts and drawings documented here suggest the goals and scope of the installation Larner envisions. More than *Wall Scratcher, Machine* will be intensely interactive. But viewers will not necessarily be aware of cause and effect in their interactions with *Machine*—at least not initially. By entering the machine's infrared and microwave "sensing field," they will trigger it to act. *Machine's* computer program is nuanced with poetic names like "Squish" and "Feeble." Emotional associations may be aroused by its shimmies, or by its erratic movements, which will intensify as the presence of viewers in the sensing field continues. Perhaps knowing that they have somehow violated the machine's "private space," viewers will retreat. Or perhaps they will persist, their presence triggering "Feeble," the programmed equivalent of a last resort—in which *Machine* tries haplessly to push itself completely out of the sensing field. In either case, Larner will have demonstrated the intricate relationships of subservience and control that we enact with our machines.

Larner's drawings for *Machine* appear, to Ana de Azcárate, as "baroque," particularly in comparison with the earlier *Scratcher.* But Angela Bulloch's *Pushmepullme Drawing Machine* returns us to the Minimalist vocabulary of Larner's earlier machine. In Bulloch's engaging piece, both the device and its product are the very picture of modest workmanship and authorial restraint. The apparatus itself looks like the flayed interior of an Etch-a-Sketch, the simple children's toy in which twin knobs control the vertical and horizontal movements of a moving point that marks the inner surface of a silvery screen. Here the marker that made its first gyrating appearance in Tinguely's Meta-matics stars again as the unlikely actor in this much more meditative drawing machine. The results of the device's patient delineations echo the important wall drawings of Minimalist and Conceptual artist Sol LeWitt, who produces simple instructions for his drawings, rather than executing them himself. The elegance lies in the instructions' simplicity (such as "lines in four colors in four directions, one millimeter apart") and in LeWitt's authorial modesty, knowing as he does that the ephemeral results will differ markedly from place to place, and from one draftsman to another. Nonetheless, the drawings always remain steadfastly "LeWitts," and Bulloch seems to make a certain sly joke out of one of the Conceptual master's most famous dicta: "The idea becomes the machine that makes the art."[38]

This literalization of LeWitt's statement, which was highly provocative when he made it in 1967, is but one of the tweaks that Bulloch gives to Minimalism's high seriousness. Another is the whimsical title, which refers to the "pushmi-pullyu," the famous beast in the children's tale *Doctor Doolittle* (played in the film version by two llamas strapped end-to-end).[39] But unlike Hugh Lofting's schizophrenic animal, whose very name suggests its deep identity problem, Bulloch's machine is quite happily itself. Do what you will with me, it announces, push or pull me with the pedal offered at your feet; your will is my desire.

As do Rebecca Horn's and Liz Larner's painting machines, Bulloch's *Pushmepullme* leaves us with many rich questions. Is the "art" the machine's proud product—the splashed canvas / scratched wall / linear mural? Or is it the endless and unceasing *process* of art-making itself, (manifestly collaborative in Larner and Bulloch alike)? Is the "artist" the maker of the machine, the machine itself, the viewer who activates it, the curator who positions it, or some collective entity? Is the medium the paint / metal / marker, or is it the medium of electricity itself, that invisible power grid which brings us these images and effects?

In *Painting Machines*, the industrial apparatus appears both in its historical specificity and as an ever-ready stand-in for our complicated selves. The machine in these works critiques traditional views of art-making, and opens up a whole new territory for exploring cultural ideas of what art, or artists, might be. Inevitably, in our postmodern frame, these paintings and machines are acutely self-conscious. They capture evanescent poetry and complex ambivalence, but they do so within an intensely knowing *matrix* (as Tansey has pithily termed it), manipulating art-historical traditions, complex technological materials, and overlapping discursive frames. The old French epithet "stupid like a painter," if it ever held truth, can hold none here. These artists understand the need for alert intelligence. They also know when to summon our ignorance, and when to hide their mechanisms in the aesthetic equivalent of engineering's "black box."

Notes

I would like to thank all those who helped me thrash out the ideas in this essay, and to convert the raw notions I began with into exhibition and text. First and foremost, my colleague Kim Sichel, historian of photography and early modernism, helped me at every point and at every level. Her tactful interventions guided my thinking and writing; she also ably steered this exhibition into port together with John Stomberg, whose humor, wit, and enthusiasm made it all fun. It is appropriate here to talk about the major contributions of my students, both in and outside of the curatorial seminar I conducted in the fall of 1995. Graduate and undergraduate students conducted research on the assigned artists, argued productively with me about my definitional categories, and, in addition to the object essays in this publication, wrote their own excellent research papers on topics relating to the project's general themes. Ana de Azcárate, Anthe Constantinidou, Ray Garraffa, Leslie Goldman, Karen Gramm, Alice Kim, Kyung-Jin Rhee, Renato Rodrigues da Silva, and Isabelle Sobin each contributed immeasurably to the working out of the ideas in this essay. I have cited their research frequently in these notes, and I hope it can be understood that the collaborative achievements of the seminar as a whole are due to all of them. My gratitude is immense. Finally, I would like to thank Peter Galison, whose comments on this essay were instrumental, and Sam Galison, whose endless inventions continue to inspire my own infatuation with machines.

1. Donna Haraway has usefully problematized these fantasies, which are given powerful form in the fictions of William Gibson (e.g. *Neuromancer* [New York: Ace, 1984]). When Gibson constructs a future in which prostitution is "a meat thing" that "happens" to be done by women, and the strangely disembodied cyberjocks are exclusively men, he is replicating standard constructs in academic (and other) discourses, in which questions about "the body" are asked by feminism, against masculinist attempts to erase actual bodies and their experiences from history. Haraway sees the talk of "disembodiedness" in virtual reality as an ideological fiction meant to deny all the living bodies involved, from the (female) lineworkers in Silicon Valley to the (female) keyboard operators risking carpal tunnel syndrome throughout the institutions of late-twentieth-century industry. I have benefited from my conversations with Haraway on the subject. See also Haraway's essay "Deanimations: Maps and Portraits of Life Itself," in Caroline A. Jones and Peter L. Galison, eds., *Picturing Science / Producing Art* (New York: Routledge, forthcoming).
2. On the origins of cybernetics during World War II, see Peter Galison, "The Ontology of the Enemy: Norbert Wiener and the Origins of Cybernetics," *Critical Inquiry* 21, no. 1 (autumn 1994): 228–66.
3. See Calvin Tomkins, *Off the Wall: Robert Rauschenberg and the Art World of Our Time* (New York: Penguin Books / Doubleday, 1980): 163–4.
4. Julien Offray de La Mettrie, *L'homme machine* (Leyden: 1748), translated as *Man a Machine* by Gertrude Carman Bussey et al. (La Salle, Illinois: Open Court, 1912).
5. The phrase "the ghost in the machine," if not originated by Gilbert Ryle, was certainly made notorious by him: "The representation of a person as a ghost [mind] mysteriously ensconced in a machine [body] derives from . . . the Cartesian category-mistake." See his 1949 work *The Concept of Mind* (Chicago: University of Chicago Press, 1984), 18. Ryle's argument is that mental processes and physical processes are in two entirely separate logical categories (as, for example, "coming home in a flood of tears" and "coming home in a taxi"). To join them with a conjunction ("she came home in a flood of tears and a taxi") is to tell a joke, and to misunderstand that "mind" and "mental processes" are in a different category altogether from "body" and "physical processes."
6. "Their shape is not the same. The base of the woman's thermometer looks like a male jewel about eight thumbs long and with a diameter of one thumb and a half; and that of the males' thermometers like the upper part of a flask which would have exactly the same dimensions in a concave sense. There are . . . these ingenious machines whose effects you will see soon." Denis Diderot, *Les bijoux indiscrets,* in *Oeuvres* (Paris: Gallimard / La Pléiade, 1962), 52. Cited by Gilbert Lascault, "Le meccaniche . . . / Mechanisms . . . ," in *Le macchine celibi / The Bachelor Machines,* eds. Jean Clair and Harald Szeemann (New York: Rizzoli, 1975), 116. Cited hereafter as *Le macchine celibi.*
7. See Pontus Hultén's unreferenced narrative in *The Machine as Seen at the End of the Mechanical Age* (New York: Museum of Modern Art, 1968), 9. For a more detailed account, see Stephen Gaukroger, *Descartes: An Intellectual Biography* (Oxford: Clarendon Press, 1995), 1, 418 n. 1. Gaukroger tells the story of Albertus

Magnus; supposedly his robot was destroyed by Thomas Aquinas "when he came across it unexpectedly in the night" (418 n. 1).

8. For American variations on this theme of machines and sublimity, see Leo Marx, *The Machine in the Garden* (New York: Oxford University Press, 1964). The texts stimulated by Marx's accounts are numerous; for a recent summary and extension of the literature, see David E. Nye, *American Technological Sublime* (Cambridge: MIT Press, 1994).
9. See my essay "The Sex of the Machine: Mechanomorphic Art, New Women, and Francis Picabia's Neurasthenic Cure" in *Picturing Science / Producing Art.*
10. Recalling that Freud termed the psyche an "apparat," some have examined Carrouges's structure and analogized it to the Freudian hierarchies of the super-ego and ego, with the bottom register of the id (repository of our animal instincts) missing from the hierarchy of these narratives. See Alain Montesse, "Lovely Rita, Meter Maid," in *Le macchine celibi*, 110-4.
11. Michel Carrouges, "Istruzioni per l'uso / Directions for Use," in *Le macchine celebi*, 21-49. For the psychoanalytic structuring of Carrouges's argument, see also the essays by Michel de Certeau and Alain Montesse, 83-97 and 110-4, respectively.
12. Indeed, as Marc Le Bot notes, Carrouges becomes a primary former of the myth he seeks to describe, and thus parallel in importance to the originating artists. See Le Bot, "Mito della macchina / The Myth of the Machine," in *Le macchine celibi*, 172-3.
13. For an in-depth analysis of Duchamp's work as it relates to issues of gender, see Amelia G. Jones, *Postmodernism and the En-gendering of Marcel Duchamp* (Cambridge: Cambridge University Press, 1994).
14. For a more extensive discussion of this phenomenon, see my essay "The Sex of the Machine" in *Picturing Science / Producing Art.*
15. In Roussel's method, which he described in painstaking detail in a posthumously published book *Comment j'ai écrit certains de mes livres*, 1930-33 (Paris: J. J. Pauvert, 1963), a simple phrase is altered by the substitution of a word or a single letter, and the gap between the resulting two meanings must be bridged by the most concise narrative possible. For example, the phrase with which Roussel begins one of his novels is as follows: "Les lettres du blanc sur les bandes du vieux billard" (chalk letters [written] at the edge of an old billiard table) is changed by the substitution of the word *pillard* for *billard*. The phrase then becomes "Les lettres du blanc sur les bandes du vieux pillard," (letters [written] by the white man about the gang of the old robber), which closes the novel. The tale in between must connect the cryptic white message with the robber's gang, and here Roussel's maniacal genius takes charge. The painting machine in *Locus Solus* is the result of a similar substitution, where "demoiselle à prétendent" becomes, by removing the "p" and pronouncing the phrase aloud, "demoiselle à reître en dents."
16. See my essay "The Sex of the Machine" in *Picturing Science / Producing Art.*
17. Jarry, *Oeuvres complètes*, vol. 1 (Paris: Gallimard / La Plèiade, 1972): 714ff, cited by Lascault in *Le macchine celibi*, 121.
18. I develop these categories of analysis in some depth in my *Machine in the Studio: Constructing the Postwar American Artist* (Chicago: University of Chicago Press, 1996). See also Kim Sichel's title essay in *From Icon to Irony: German and American Industrial Photography* (Boston: Boston University Art Gallery, 1995).
19. Exemplifying this shift was the invention of the video casette recorder, developed by American engineers and manufactured exclusively by Japanese corporations. This move marked the end of a long tradition in which American patents had dominated the world, and where the United States' perfection and production of other countries' preliminary designs had been the sign of its vigorous "can do" economy.
20. Henry Adams, *The Education of Henry Adams*, 1906 (Boston: Houghton Mifflin, 1946), 379–90.
21. Obviously, photographic practices were already being modeled on these industrial approaches.
22. See Louis Kaplan, "The Telephone Paintings: Hanging Up Moholy," *Leonardo* 26, no. 2 (1993): 165–8.
23. It seems important to stress these examples, for they illustrate so clearly that the American models of Taylor and Ford were seen initially as purely progressive, and not inherently capitalist, at least in the Soviet (Russian) and Socialist (Bauhaus) interpretations. See my *Machine in the Studio* for further discussion. For more on Taylorism and Fordism, see Thomas Hughes, *American Genesis: A Century of Invention and Technological Enthusiasm* (New York: Viking Penguin, 1989).
24. It was far easier, for example, for Schlemmer to "automate" the human body and denaturalize its movements with padded costumes and geometric choreography than for him to accomplish his much more ambitious goal (out of reach in both financial and technological terms): a totally mechanized theater of light, sound, space, and shape (still operated, of course, by the "man in the control booth"). And although the artists of revolutionary Russia tried to merge their creative impulses with the ethos of the assembly line, it was still up to a handful of charismatic individuals to stamp the moment with their unreproducible and impractical vision. Thus the single most powerful image of that time remains Vladimir Tatlin's *Study for a Monument to the Third International*. This populist structure, which exists only in plan and photographs of a maquette, reconfigured

the Tower of Babel as a radio antenna, tilting and thrusting dynamically upward toward a slender point. While Tatlin's associates could read this at the time as a statement of collectively industrializing community (as Lenin said, "Communism is the Soviet people plus electricity,") we are more likely to read it today as a figure for utopian individualism thwarted by an ossifying totalitarian regime. For research on the theater of the Bauhaus, I have relied upon the unpublished paper by Alice Kim, "The Bauhaus Theater Workshop: From Men to Marionettes and Automatons." Information on Soviet Constructivism and Productivism can be found in the essays for *The Great Utopia: The Russian and Soviet Avant-Garde*, 1915–1932 (New York: Solomon R. Guggenheim Museum, 1992).

25. A thought experiment helps convey what I mean. If octopuses produced technology, handles would be orbs, not cylinders. In place of screws and bolts, which humans are incapable of perceiving outside a system of sexual signification, one might have varieties based on the octopoid suction cup (which respects no sexual division). One can quickly see how humans' bipedal symmetries, sexual dimorphism, and planar-pincer grasp have each marked the supposedly neutral terrain of technological design.
26. For an analysis of Francis Picabia as one such exception, see my essay "The Sex of the Machine" in *Picturing Science / Producing Art*.
27. For an analysis of this period in terms of visual art, see my *Machine in the Studio* (see note 18 above).
28. Tinguely via Pontus Hultén, quoted in "Some Authors of Bachelor Machines," *Le macchine celibi*, 217.
29. In one trajectory, performance became everything, and there is a mechanolatrous realm of this type of art as well. Laurie Anderson is perhaps the best known example of those artists who use technology as an integral part of their performance art. Whether she is threading magnetic tape into her violin bow and wiring tape heads onto her violin, or becoming one with her vocoder, she has an extraordinary capacity to use machines to comment on mechanolatry itself.
30. Henry Luce, "The American Century," *Life,* February 17, 1941; reprinted in John Jessup, *The Ideas of Henry Luce* (New York: Atheneum, 1969): 106–20.
31. I hasten to insist, however, that *all* present-day art forms are "technological," as Duchamp reminded us when he pointed out that the tube of oil paint was a readymade. I would even go so far as to suggest that the Neolithic cave-painters were technologists, if we interpret *techne* to include the use of sticks, fire, and stones to gouge, burn, and grind pigments.
32. In the history of twentieth-century art, there have been many women painters of industrial environments or mechanomorphic forms: the American Precisionists Elsie Driggs and Georgia O'Keeffe, Dadaist Hannah Höch, Surrealists Kay Sage and Leonora Carrington. Female photographers of the machine are also legion—from Germaine Krull with her extraordinary portfolio *Métal*; to Berenice Abbott and her urban panoramas and Margaret Bourke-White, with her industrial scenes; to contemporary interrogators of the human/industry interface, such as Martha Rosler and Deborah Bright. My thanks to Alan Helms for drawing Deborah Bright's work to my attention. See Grant H. Kester, "Swept Away: Deborah Bright and the Fossils of Capitalism," in *All That Is Solid* (Atlanta: Atlanta College of Art Gallery, 1997).
33. I thank Ana de Azcárate for bringing this installation to my attention. Hirst provided paints and paper on adjacent pedestals so that visitors could choose their own combinations to make paintings (as I would call them, rather than drawings) to join the framed ones (by Hirst) on the gallery's walls. This project, clearly analogous to Tinguely's 1959 Meta-matics shows, represented something of a fluke in Hirst's production, fitting generally within his explorations of the boundaries of taste (here courting kitsch, but elsewhere seeking the effects of horror or disgust), rather than constituting part of any ongoing investigation of technology. Mark Pauline of San Francisco's Survival Research Labs (SRL) also arguably makes "painting machines," small or gigantic mobile mechanoids that spew fluids as they careen through space. Unlike Hirst's installation, however, SRL speaks not to the conventions of art but to other social gatherings such as demolition derbies and punk spectacles.
34. Debra Bricker Balken, "'Deus ex machina': The Work of Medrie MacPhee," in *Medrie MacPhee* (Milan and New York: Paolo Baldacci Gallery, 1995), 7. I am grateful to Marta Braun for bringing MacPhee's work to my attention; unfortunately, work on the current exhibition had progressed too far to include any of MacPhee's evocative recent work. In any case, MacPhee's paintings are different from the canvases shown here, for she draws directly on the vocabulary of early modernists' mechanomorphic art to evoke a similarly uncanny play between the machine's metallic surfaces and the body's more permeable form.
35. See Laura Billings, "Chip Chat: Drifting in Cyberspace," *Family Life,* November 1995, 66.
36. Guy Debord, *The Society of the Spectacle*, 1967, (Detroit: Black and Red, 1977).
37. This "trace" can be connected to the philosophical writings of Jacques Derrida, who has been an influence on Mark Tansey. See Derrida's *Margins of Philosophy* (Chicago: University of Chicago Press, 1982), 65.
38. Sol LeWitt, "Paragraphs on Conceptual Art," *Artforum* 5, no. 10 (summer 1967): 79–83; excerpted in Charles Harrison and Paul Wood, eds., *Art in Theory, 1900-1990* (Oxford: Blackwell Publishers, 1992), 834.
39. The pushmi-pullyu is "the only two-headed animal in the world." See Hugh Lofting, *The Voyages of Doctor Doolittle*, 1922 (New York: Lippincott, 1950), 61.

THE ARTISTS

Angela Bulloch

Lawrence Gipe

Rebecca Horn

Liz Larner

Robert Moskowitz

Donald Sultan

Mark Tansey

Rosemarie Trockel

ANGELA BULLOCH

Born in Ontario, Canada, 1966. Lives in London.

British-based artist Angela Bulloch has created many installations of her drawing machines, including *Blue Horizons* (1990), *On/Off Line Machine* (1990), and *Pushmepullme Drawing Machine* (1991). Time, mechanization, and the potential for viewer participation in the work of art define her "painting" devices. With simple mechanical systems and the use of the wall as a drawing surface, *Pushmepullme* (page 33) invites the viewer's active manipulation of the work. The gallery wall acts as a canvas, and the simple mechanized arm holding a pen acts as the artist. But this "robot-artist" needs direction. Viewers can, if they choose, assume the "artist" position, effecting change by controlling, with a foot pedal, the speed and direction the mechanical arm will take. This is not a passive work of art stretched behind a wood frame; rather it is an active, evolving artwork with metal pieces moving across its surface—constantly altering and defacing the wall. In fact, the wall, which usually acts as a submissive host to paintings, with its neutral background color, becomes an active agent of dialogue.

Pushmepullme poses many intriguing questions about art and technology, artist and viewer, media and medium. The theme of *time* emerges in the evolution and metamorphosis of the "painting." In the beginning only a few red lines appear on the wall, but eventually the pen's ceaseless motion and repetition render a mass of networked lines that nearly cover the surface with a mesh of red. The speed of the mechanized arm resembles a wildly random metronome, evoking the passage of time, but only as inflected by the viewer at the helm.

In reference to her viewer-activated works Bulloch has commented, "The renegotiation of power interested me."[1] This renegotiation is suggested in the title of the piece. Who is the "me" of *Pushmepullme*—the artist, the viewer, the wall, the pen, or the machine? Not only is Bulloch interested in renegotiating power through the art she creates, but she also compels the viewer to do the same. Reinterpretations and shifts in the roles of art and viewer are fundamental components of Bulloch's work.

—Isabelle Sobin

1. From an interview with Francesco Bonami, "Angela Bulloch: Lonesome Comedy," *Flash Art* March/April 1992, 96–7.

Angela Bulloch

Pushmepullme Drawing Machine, 1991

Steel armature, electric motor, felt-tip marker, chair, and foot switch

120 x 360 in.

Private Collection, Courtesy Deitch Projects, New York

LAWRENCE GIPE

Born in Baltimore, Maryland, 1962. Lives in California.

The paintings of Lawrence Gipe often invoke the visual language of the sublime while challenging us to examine our notions of progress and industrial power. Gipe accomplishes this by choosing imagery directly from the pro-industry propaganda found in magazines and advertisements of the golden age of the machine, the 1930s and 1940s. He challenges the messages of these images through the use of text superimposed against their inviting, powerful depths.

Painting #4: Themes for a fin-de-siècle (page 22) and the *Triptych from the Krupp Project* (page 36) both form part of the project entitled *Themes for a fin-de-siècle* (Themes for the End of the Century). This project encompasses *The Century of Progress Museum, The Krupp Project,* and other cycles of paintings that examine the mingled emotions of optimism and apprehension that were felt at the beginning of this century. All these paintings bear one of seven words: Desire, Greed, Complicity, Order, Pride, Faith, and Acquisition. Gipe has stated, "I thought that these words poetically described the century in some way and symbolized progress. . . . These words then became headings under which related subjects could then be examined."[1] Gipe's seven words within these cycles of paintings also suggest the seven deadly sins.

In *Painting #4: Themes for a fin-de-siècle* he presents us with an industrial landscape dominated by the silhouette of a bridge heroically spanning a body of water. The scene is illuminated from behind by the sun, its light filtering through the emissions of a factory, creating a smoky, colorful sky. The atmospheric effects in sky and water draw us into the image, even as we become aware that their beauty may actually be the result of chemical pollution.

Emblazoned across the hazy landscape, in bold red letters, is the word *Pride;* the bridge seems to echo the word, straddling the river like a triumphal arch. To our late-twentieth-century eyes, the use of this word in conjunction with the implied destruction of the environment seems rather ironic. In our culture of recycling, the Clean Air Act, and smog warnings, the release of pollutants in the name of progress does not evoke a feeling of pride. In this image, we are looking into the past through the present, contemplating the result of a misguided emphasis on progress.

In the *Krupp Project* triptych, Gipe moves from the general condemnation of "Industry" as a nameless and timeless force, to the depiction of industrial greed within a specific historical context. The Krupp family made its fortune in Germany (in railroad construction, steel, and armaments manufacture) in the nineteenth and twentieth centuries. Alfred Krupp was deeply involved with the Nazi military effort, staffing his factories with slave laborers from the concentration camps. When workers no longer met production quotas, they were transported to their deaths. At the end of the war, Krupp was sentenced to ten years in prison but he served only thirty months; the United States believed he would be instrumental in the cold war by helping with the economic development of postwar Germany, and thus aiding in the fight against Soviet influence in central Europe. "Complicity" thus refers both to Krupp's American protectors and to his German colleagues-in-arms.

Despite its dark historical theme, the *Krupp Project* triptych is full of light effects that suggest the presence of the divine in the industrial world. On the right panel of this "altarpiece," we see the factory in all its glory, accompanied by a text taken from slogans used by the Krupp company to motivate their workers. The words are in praise of the company: "Gott segne das Haus und die Firma Krupp wie bisher so auch in alle Zukunft" ("God bless the house and the firm of Krupp, now and in the future"). On the triptych's left-hand side, we actually enter into the world of industry and are surrounded by it, rather than simply observing it from the outside. The text here reads: "Zum teil der Werksangehörigen und des ganzen deutschen Volkes" ("Part of the workforce and all the German population"). Rays of light illuminate the factory in the same way that light enters a cathedral through stained glass windows. In the middle panel, we see the gilded statue of Atlas from New York's Rockefeller Center, an icon associated with the United States financial capital. Planes fly overhead and a skyscraper looms in the background. This imagery evokes a world of overwhelming economic power, made masculine in Atlas's muscle-bound form. While the image glorifies the United States by linking it to such power, the label *Complicity* suggests that the government ultimately allied its power with Krupp's, rather than working against it, as wartime propaganda had claimed.

Lawrence Gipe uses seductive imagery, coupled with ironic text, to denounce the power of industry and progress. The religious connotations of the "seven words" (the sin of "pride" chief among them), together with the exalted imagery of rays of light, further reinforce a connection between the the power of the medieval European Church and the power that industry wields in the twentieth century. Gipe warns us against the danger of such power in anyone's hands.

—Leslie Goldman

1. Trinkett Clark. *Parameters* (Norfolk, Virginia: The Chrysler Museum, 1994), 2.

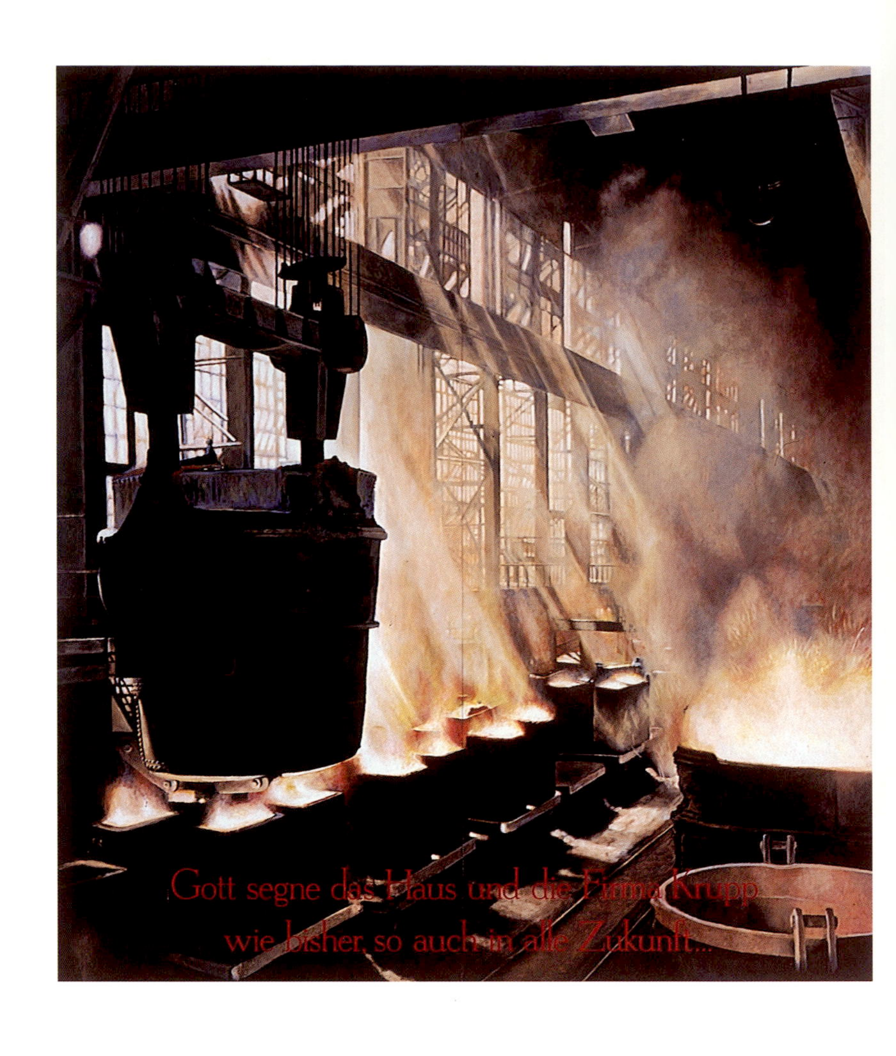
Gott segne das Haus und die Firma Krupp
wie bisher, so auch in alle Zukunft...

Lawrence Gipe
Triptych from the Krupp Project, 1990
Oil on wood panel
Central panel—90 x 70 in.; side panels—90 x 80 in. each
Los Angeles County Museum of Art, Gift of Barbara Schwartz in Memory of Eugene M. Schwartz

REBECCA HORN

Born in West Germany, 1944. Lives in Berlin.

Kleine Malschule (Little Painting School—page 39) exemplifies the type of sardonic creativity characteristic of the German-born artist Rebecca Horn. The work initially elicits a comparison with the process-oriented gestural Action Paintings of Jackson Pollock and other New York School artists; its very title suggests this parallel. Horn, however, emphasizes the process of painting rather than the painting itself and, most importantly, substitutes a machine for the body of the artist. That a female artist produced this work implies a commentary on the rhetoric of gestural painting, a male-dominated movement (as was Surrealism, whose influence is also present in Horn's work); thus Horn challenges the role of gender in art history, besides challenging the conventional art of painting. She has worked with everything from fiberglass, metal, and polyester, to pigments, earth, and feathers—which indicates her lack of interest in traditional painting and suggests a subtler critique of genre and of the artist's role.

A dominant theme in Horn's work—springing from the artist's desire for communication—has been the construction of human extensions. Emerging from her performance art and film, these animal/artificial extensions were first intended to be worn, as the soft, life-sized, feathered wings were worn by the nude woman in *Paradise Widow* (1975). As her work progressed, these extensions began to exist on their own, equipped with motors that allowed them to perform with no apparent human intervention. By 1985, when she created *Brush Wings,* the soft paintbrush "wings" had been attached to a mechanical sculpture rather than a human body, the delicate fan of brushes opening and closing in a mechanically timed sequence. With *Kleine Malschule,* this mechanical "freedom" becomes absolute, within the limits of nonobjective gestural painting. Horn's machine performs the process of creating a work of art, both enacting and concealing her own creative activity in constructing the machine in the first place. She has created a surrogate artist.

A Mary Shelley of the twentieth century, Rebecca Horn is an artist in the truly magical sense that she gives life to her creations. Reinterpreting Shelley through a postmodern lens, Horn illuminates the complexities of the role of Frankenstein as both creation and creator. She gives life to otherwise inanimate objects—not only by installing motors that cause them to "faint" and "die" and "come back to life again"—but also by imbuing them with human attributes. The poetic soul she gives these machines calls forth an emotional response from the audience, a response that may account for so many comparisons of her work to "medieval alchemy."[1]

—Alice Kim

1. See, for example, John Dornberg, "Rebecca Horn: The Alchemist's Tale," *Art News* 90 (December 1994): 94–9.

Rebecca Horn

Kleine Malschule (Little Painting School), 1988

Metal construction, ladles, brushes, motor, wood stretchers, and pigment

116 x 17 x 43 in.

Private Collection Switzerland, Courtesy London Projects

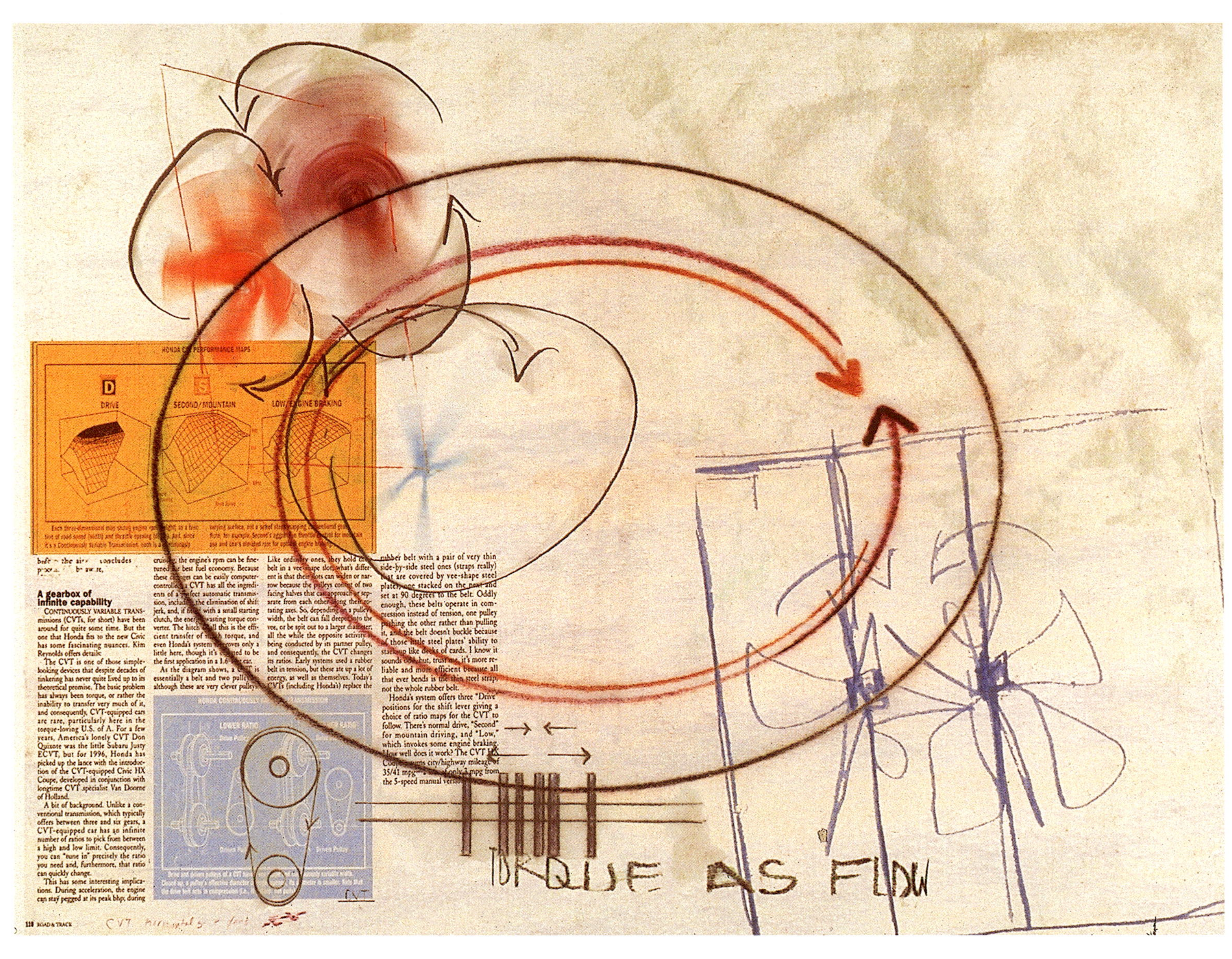

Liz Larner

Fans from *Machine Drawings*, 1996–7

Mixed-media Iris print collage

20 x 24 in.

Courtesy of the artist

LIZ LARNER

Born in Sacramento, CA, 1960. Lives in Los Angeles.

***W**all Scratcher* (page 43) is a compact machine with an arm that moves continuously to scratch a gallery or museum wall with a heavy-duty steel needle. It incises the wall in a repetitive manner that somehow blends refinement with violence, not unlike a fingertip that is stroking, or picking, in an obsessive way.

The sculpture is repositioned within the gallery on a daily basis and thus leaves marks in different places. These marks have been compared, by the critic David Pagel, to the scratches "made by prisoners to record the length of their incarceration."[1] Thus *Wall Scratcher* can be said to be commenting not only on the repetitive and cyclical nature of machine motion, but also on the repetitiveness of day-to-day life itself.

Like Larner's more violent *Corner Basher,* also from 1988 (composed of a steel ball-and-chain that repeatedly smashes a wall), *Wall Scratcher* seems to embody the artist's urge to destroy the walls of a gallery, as a way to free the work of art from arbitrary institutional confines. Yet it also expresses Larner's interest in touch and gesture, linking her work to that of Rebecca Horn, whose machines are a fundamental source for Larner's artistic inspiration. Horn's projects from 1982 include circle and radius inscribers that repeatedly mark the gallery's floor and wall, the former with a felt-tip marker, the latter with a needle. Both artists' mechanical devices must be in constant motion to perform their work—movement thus becoming an expression of the powerful and transgressive potential of art.

The *Machine Drawings* (pages 40, 58–61) reflect Larner's initial attempts to draft plans for a complex interactive installation: an apparatus designed to act, to move once a person penetrates its sensing field, the circular area surrounding it. The activity of the *Machine* depends on the intensity of heat and motion within this sensing field.

At present, the drawings are all that exist of *Machine;* within them it is being gestated. Its movements are indicated by the interrelation of geometric figures and by Larner's spontaneous, rapid marks on paper; through a static medium the artist expresses the machine's potential for movement. Compared with *Wall Scratcher* and *Corner Basher* these drawings are complex, almost baroque—a collage of geometric figures in earthen colors, with handwritten and typeset text explaining *Machine's* purpose, and clippings explaining the technologies on which it is based. The dense drawings contrast with Larner's naked Minimalist sculptures.

The complexity of the drawings appropriately reflects the plans for this enormous and complicated device that will perform a number of actions at once. It fills a space, surrounds it with action, the whole room becoming its acting field; *Wall Scratcher,* in contrast, makes only a minimal intervention in space.

Larner's basic impulse is not so much to deconstruct the machine, as to make an active apparatus that expresses emotions through its movement, movement that is linked to interactivity: *Machine* will act only when a living organism penetrates its sensing field.

Machine can be compared to Jean Tinguely's interactive Meta-matics, drawing machines from 1959. To use one of Tinguely's machines, the viewer inserted a piece of paper, clamped it onto a metal support, and secured a writing utensil in an arm-wire; the Meta-matic was then switched on and the arm started to move across the paper, which was also in motion. The machine's restless movements resulted in an abstract gestural drawing.

Machine and the Meta-matics both need human participation to start their performance. For the former, interactivity is crucial; with Tinguely's Meta-matics, however, the relationship of machine and person was rather superficial. In Larner's work there is intimacy and dependency between machine and viewer; the observer becomes part of the work itself.

Liz Larner has always been interested in movement and change within her work. In her early "Petri dish" pieces such as *Orchid, Buttermilk, Penny* (1987), she incorporated fluids and objects into growing medium, producing different reactions that continuously changed the piece. In contrast to *Machine*, which will perform its movement through electronics and electrical mechanisms, these earlier pieces were shaped by biological systems that exhaust themselves and end in decay. The machine could end up deranged, the artist says, but it will never break down. The implication is that death awaits the organic, but the mechanism never dies.

—Ana de Azcárate

1. David Pagel, "Liz Larner," *Arts Magazine* 63 (December 1988): 93.

Liz Larner

Wall Scratcher, 1988

Anodized aluminum, spring steel, 12 volt motor and battery

47 x 18 1/8 x 12 1/4 in.

Collection: Ira and Lori Young, West Vancouver, Canada

ROBERT MOSKOWITZ

Born in New York City, 1939. Lives in Manhattan.

Since the beginning of his career, in 1959, Robert Moskowitz has been making a great deal of effort to reconstruct painting. Part of a generation that was transgressing the traditional boundaries of the medium (a generation that included Jasper Johns, Robert Rauschenberg, Frank Stella, and Andy Warhol), Moskowitz was interested in discovering the nature and principles of painting. In a period defined by radical experimentalism—when the basic attributes of an artistic tradition were bluntly denied—he went against the grain, laboring to refine and perfect pictorial conventions. Today, his unique contribution presents a set of accomplishments that are currently undervalued in histories of the period.

In his effort to achieve the reconstruction of painting, he first investigated the materiality of the canvas in a series of collages; he later studied systems for the compositional notation of the figure, proceeding finally to examine what produces meaning in a painting. In a series that began to emerge in the 1970s, Moskowitz makes reference to Mondrian, Rodin, Brancusi, and Giacometti, establishing the limits of his investigation between the poles of abstraction and representation. Quoting a modernist icon such as Mondrian's *Red Mill,* Moskowitz distills the subject of the painting to focus on the frontality and geometry of the central windmill form. When he quotes Rodin's *Thinker* he conceals its Impressionist surface in an enigmatic silence. Thus Moskowitz's icons stage a dialogue between the subject, rendered by modern artists in all its complexity, and the subject's distilled artistic and symbolic features. The same method is applied in his landscapes and cityscapes of the 1980s.

Stack (page 45) characterizes this phase of Moskowitz's artistic maturity. Here, in place of distilled monuments of high art, he presents another kind of modernist icon—the slightly tapered column of a factory smokestack. The implied equations between these icons in Moskowitz's work are intriguing. Both windmill and smokestack, after all, are industrial forms—but the former has been aestheticized, made geometric and iconic by no less an artist than Mondrian (and before him Jacob van Ruisdael, and Rembrandt, in the best Dutch tradition). The smokestack, too, has been the subject of artistic fascination; there is a 1925 photograph of factory stacks taken by Charles Sheeler that directly influenced Moskowitz's composition. One also thinks of Giorgio de Chirico's "metaphysical landscapes" in which factory smokestacks cast long, elegiac shadows over abandoned urban squares. Unlike Sheeler or de Chirico, however, Moskowitz has isolated and dramatically enlarged the scale of his *Stack* by compressing it within an enormous vertical canvas. The factory's most aggressive visual feature is selected, and its single smokestack stands in for the implied whole of the factory environment. The obvious sexual connotation of the motif enhances its oppressive power—yet this is manifestly an *inactive* smokestack; no smoke belches forth, and no sparks fly. By implication, the underlying infrastructure that feeds the stack is also still. This is thus a postindustrial image that suggests mourning and loss.

Robert Moskowitz
Stack, 1979
Oil on canvas
108 x 34 1/2 in.
Private collection

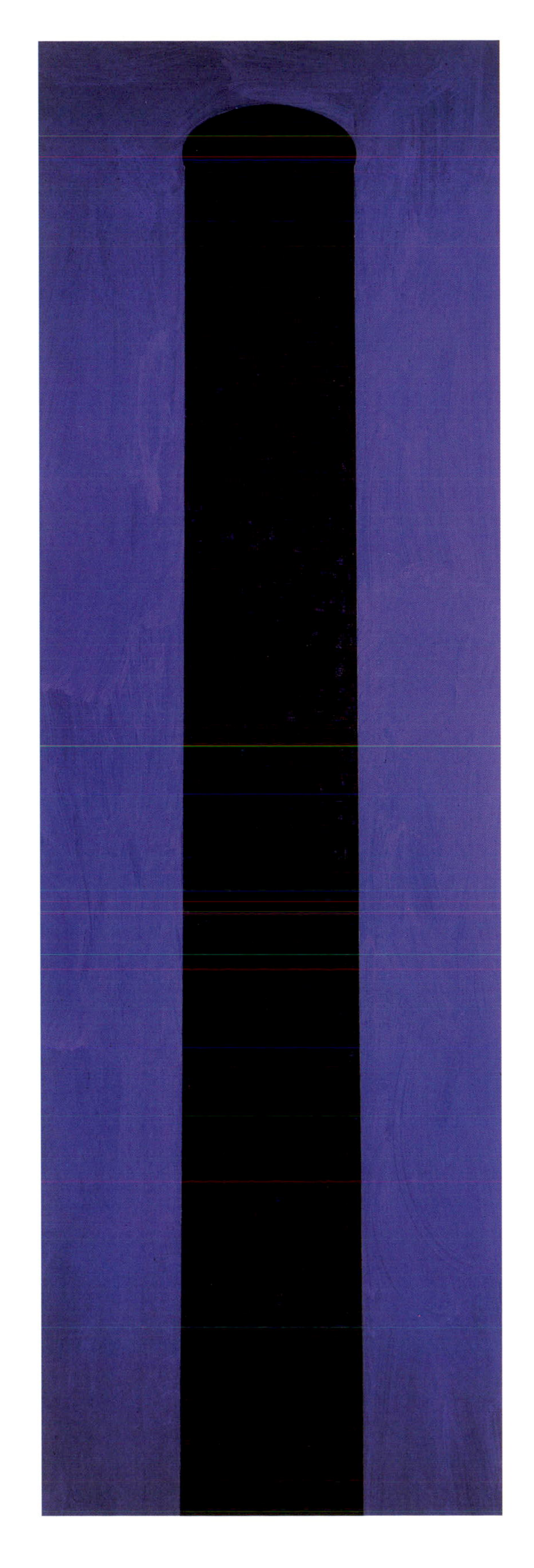

In this and other images of the stack, Moskowitz works to obscure the recognition of his motif. As he deliberately confounds figure (the blackened stack) and ground (close-valued, flatly painted, blue-gray surround), the artist disturbs our normal patterns of visual perception. We ordinarily rely on clear "gestalt" relationships between small dark silhouettes and the larger light fields that surround them, much like the relationship between these black letters and this white page. As Moskowitz himself has noted, "They [figure and ground] can easily slip back and forth for me, and it isn't as simple as the image occupying the foreground space."[1] This ambiguity, added to the image's references to industry and masculinity—both eerily emptied of any active life—give *Stack* a powerful, even terrifying presence.

—Renato Rodrigues da Silva with C. A. J.

1. See Ned Rifkin, *Robert Moskowitz* (New York: Hirshhorn Museum and Sculpture Garden / Thames and Hudson, 1989).

Donald Sultan

Plant, May 29, 1985, 1985

Latex, tar, and fabric on vinyl tile over masonite

96 x 96 in.

Hirshhorn Museum and Sculpture Garden, Smithsonian Institution. Thomas M. Evans, Jerome L. Greene, Joseph H. Hirshhorn, and Sydney and Frances Lewis Purchase Fund, 1985

DONALD SULTAN

Born in Asheville, North Carolina, 1951. Lives in New York.

Donald Sultan draws his subjects from newspaper photographs and uses industrial materials in his work. His imposing images, usually eight feet square, comment on the industrial age. They may well emerge from his experience during his youth in North Carolina, where he worked in industry.

In the mid-'80s, Sultan began a series of paintings of machine-age disasters; depictions of airplane crashes, guerrilla bombings, and industrial fires began to dominate his work. These subjects lent themselves well to the industrial materials Sultan was then using, which required an innovative process of canvas preparation that he still employs. First he applies thick coats of tar or butyl rubber to common vinyl tiles mounted on masonite; he then "carves" into this thick surface with a knife or blowtorch. Later he builds the surface back up again with plaster, and finally paints on the mass of materials. As one critic has remarked, "Sultan enjoys the fact that his forest fires and poisonous gas fumes, his burning buildings and scorched landscapes are created by fire and smoke in the studio."[1] What at first may seem contradictory is the richness of Sultan's finished work. Though disasters are often the subject, and industrial materials his media, he manages to imbue the panels with a subtle sensuality. This contradiction is only one of many ironies that emerge from Sultan's work.

Plant, May 29, 1985 (page 46) refers to a specific event: the closing of a steel mill in Ohio. From the title alone, the viewer is likely to expect something organic; images of vernal landscapes may come to mind. Unexpectedly, however, Sultan draws on the multiple meanings of *plant,* juxtaposing an industrial landscape diametrically to the pastoral scene. Of this work Sultan has said, "Plants are dying, one of my paintings is a picture of a shut-down steel mill in Ohio. The title is *Plant:* there are barely any trees and the factories are dying."[2] By depicting this emptied factory in such an ominous way—its dysfunctional, dead "plant" form devoid of any green—Sultan seems to be condemning industry. Nevertheless by frankly using commercial materials, he simultaneously is endorsing industry; the painting seems to be mourning its loss. A final contradiction: although *Plant* is a dark scene, it is depicted with the sumptuous strokes and awe-inspiring style of eighteenth-century Romantic painting. Sultan works with his textured media to create a sensual portrait that flirts openly with the sublime.

As in *Plant,* Sultan's more recent *Polish Landscape II, January 5, 1990* (page 48) is dominated not by figure or color, but by the rhythm created by lights and darks as they resonate across a textured background. With his use of thick tar as background, Sultan creates a landscape both barren and tumultuous. The large, sweeping landscape evokes the sublimity of J. M. W. Turner. At first glance, *Polish Landscape II, January 5, 1990* does not appear to be a disaster. Nonetheless, Sultan has rendered it in much the same manner as *Plant*, confirming that he sees the two as analogous. An empty train platform becomes a memorial, recalling the hundreds of such stations in Poland used by the Nazis and their allies to trans-

Donald Sultan

Polish Landscape II, January 5, 1990, 1990

Latex and tar on vinyl tile over masonite

96 x 96 in.

The Eli Broad Family Foundation, Santa Monica

port prisoners to concentration camps. Of Russian-Jewish heritage, Sultan is personally invested in the meaning of *Polish Landscape II.*

Without background information, however, both *Plant* and *Polish Landscape* are ambiguous. Sultan wants the viewer to make judgments based not only on the information he has provided, through the work and its title, but based also on the experiences that an individual brings when observing such a painting. He believes that "a person brings to painting his own set of perceptions and it is the painting's job to bring about a continual dialogue with those perceptions—no matter at what level they may be. . . . You have to engage people and let them fill in the [meaning of the] picture."[3] For this reason, *Plant, May 29, 1985* and *Polish Landscape II, January 5, 1990* are both provocative but ultimately unresolvable works; viewers can read them as their experience dictates.

—Karen Gramm

1. Carolyn Christov-Bakargiev, "Donald Sultan," *Flash Art*, May/June 1986, 50.
2. Museum of Contemporary Art, Chicago, *Donald Sultan* (Chicago: Museum of Contemporary Art, Chicago, 1988), 25.
3. Ibid., 18.

MARK TANSEY

Born in San Jose, CA, 1949. Lives in New York.

The four works by Mark Tansey shown in this exhibition represent machines as metaphors for artistic production. In the painting and in two of the drawings, the machine is shown as the creator of art. But only when the machine is controlled by a man is the task completed successfully. When the machine is left unsupervised, the result is disastrous.

The three drawings are part of the suites created by Tansey specifically for his retrospective exhibition in 1993. In an introductory essay, he notes that the drawings were created under "a schematic notion of pictorial content based on the wheel."[1] The wheel is also found in Tansey's own "idea machine" or "content wheel," made of three concentric paper disks, laid one on top of another and connected at the center. Around the edge of each disk appear words—bits of phrases. The nouns on the outer disk constitute various sentence subjects; on the middle ring are verbs; and on the inner ring, nouns constitute objects. The wheels are turned, and the resulting sentences, formed by the combination of the words, give Tansey the ideas for his paintings.

The drawings exhibited here are taken from two series, *Wheels* and *Frameworks*. The diptych consisting of *The Hub* and *Discursive Formation* (pages 62–63) belongs to the *Wheels* suite, which focuses on the circular mixing of the content—a metaphor for Tansey's "content wheel." On the left is a wheel of nature, created by logs, water, and rocks. The central point of activity is used to create a dynamic image of motion, but there is no disturbance in the circular movement of the hub. This is changed in *Discursive Formation,* where a machine enters the center of the "wheel," creating chaos. The piece's original title was *Vigilant Machinery Caught in Discursive Formation,* which is clearly explanatory. *Vigilant* is another word for *watchful*, and discursive may be defined as "rambling from one subject to another" (it may also refer to post-structuralist theories of signification). Thus the "object" is caught in a rambling path from one "subject" to another, to form an idea that will become art. During this process, different subjects enter and exit simultaneously—a chaotic situation. Meanwhile the machine attempts to keep up with the rate of production, but Tansey suggests it is "caught," or trapped.

In *The Raw and the Framed* (page 6), Tansey uses the frame structure as a metaphor for representation—exploring, as he has said, "the idea of representational content as matters of framing: art production as framing, . . . frame as substitutes for subject matter."[2] In this drawing, he is interested in the analogy between artistic production and mass production, in depicting the production of art as a mechanical process. On the left, two miners work with machinery to chisel out fragments of rock, which are placed on a conveyor belt and transported through the frames to the next scene; somewhere along the way, the rocks (*raw* material) are transformed into *framed* paintings. At the far right, two men examine the finished products.

Mark Tansey

The Matrix, 1993

Oil on canvas

104 x 74 in.

The Eli and Edythe L. Broad Collection, Los Angeles

Tansey reiterates the idea of mechanical replication through his drawing method. After selecting an idea from his "content wheel," he uses a photocopier to combine a selection of various images from his picture file, and the resulting collage is often the basis for a graphite drawing, as in the case of *Hub*. In other instances he intervenes in the photocopying process, manually interrupting the progress of the paper through the machine and removing areas of toner. He alters the original image, and renders it unique—a process analogous to the mechanical art production in *The Raw and the Framed*. There the framing of the scenes imparts a narrative quality: from left to right, one sees the process of the manufacture and selling of art.

As he does in many of his paintings and drawings, Tansey critiques the art industry and also himself as an artist. He wants to see art-making as mechanical production, much like the *56 Brush Strokes* created by Rosemarie Trockel's *Painting Machine*. One difference, however, is that Tansey hides his mechanical assistant—the photocopier is never on view—while Trockel has often exhibited her *Malmaschine*.

Tansey's analogy between mass reproduction and art production can also be found in *The Matrix* (page 50), a painting created from an earlier drawing, in the *Frameworks* suite, entitled *The Bricoleur Redeploying the Framework* (1992). In *The Matrix,* the central apparatus controls what happens around it. In contrast with *Discursive Formation*—where the machine seems helpless, overwhelmed by the information and the combination of elements feeding into it—the machine in *The Matrix* gathers and combines different components to produce successful finished art.

The matrix partitions the painting into four sections, creating smaller frames of activity. From three sides, paint tubes, text, and raw material are fed into the central apparatus by conveyor belts, and emerge at the right as conventional framed paintings. A man, presumably the artist, controls and directs the machine. As in *The Raw and the Framed,* Tansey here considers the mass production of art. The artist, positioned in the center as the controller of production, becomes analogous to a machine that reproduces art.

—Anthe Constantinidou

---

1. Cited in Judy Freeman, *Mark Tansey* (Los Angeles: Los Angeles Museum of Art / Chronicle Books, 1993), 69.
2. Ibid., 70.

Rosemarie Trockel

Untitled (Woolmark), 1986

Machine-knitted wool,

86 1/2 x 16 in.

Marjory Jacobson and Marshall Smith, Boston

ROSEMARIE TROCKEL

Born in Schwerte, Germany, 1952. Lives in Cologne.

Rosemarie Trockel is a post–World War II, German-born conceptual artist whose work suggests the twin influences of Andy Warhol and Joseph Beuys. Her deadening repetition of images stems from Warhol, but in works like *Untitled* (*Woolmark*—page 52), the machine-knitted image is also a statement about the place of women's work in the spheres of the domestic, fashion, industrial, and art worlds. The repeated image of the international wool trademark is a comment on the corporate logotype, on machine-made products in general, and, in particular, on the relationship between a product's manufacturer and the female user. The recognizable image of woman-as-consumer, looking for the woolmark before a purchase, confronts the viewer in the context of "female" woven textile art.

The use of industrial machinery to knit these "paintings" underscores the resonance that Trockel's work has with that of Beuys, who believed that "the whole social fabric is a work of art." But while these knitted paintings manifest their status as commodities—employing the woolmark logo and recalling Warhol's use of brand names—they also bring to mind the industrialization of women's work. These complex productions call into question the "distaff side" of creativity: the realm of carding, spinning and knitting animal fibers into complex, ethnically distinct patterns—traditional women's work. More clearly manifest in Trockel's knitted masks, dresses, and "schizo-pullovers" (sweaters with two necks), there is a gap that these knitted objects serve to highlight—the gap between women's handiwork and the industrial processes brought to bear on their commodification.

Ich kannte mich nicht aus (I am Stumped or I Don't Know About Myself—page 55) is a static apparatus rather than a potentially functioning machine, suggesting a clothes-drying or wool-dyeing rack. It, too, evokes the female domestic sphere. And yet, by draping yarn loosely around the arms of the metal stand, Trockel removes that humble material from its traditional role in hand-knitting and suggests a more industrial setting.

Like a necklace, a photograph hangs from the rack; this renders the apparatus nonfunctional and adds a further reference, for the photo depicts a figure holding or manipulating a simple, hemispherically capped spring. The device is "a turn-of-the-century crackpot invention intended to demonstrate proof of the existence of God, a patent which Trockel purchased in 1988 from the U.S. government."[1] Phallic symbol or industrial mythological relic, the object is manipulated by the hands of the implied female operator of the nonfunctioning apparatus. But, as Trockel acknowledges in her title (which belies the manifest technological sophistication of her other works), such masculine machines remain incomprehensible. This mixed-media work serves again to suggest the tension between evolving male and female roles in relation to technology.

The drawings made by Trockel's *Malmaschine,* (Painting Machine, 1990, shown in figure 4, page 24), have many connotations, and force the viewer to confront the provocative concept of the total removal of the artist's hand from the process of producing art. Titled *56 Brush Strokes* (page 56), these drawings are not produced by an intricate or high-tech apparatus. The *Malmaschine* suggests instead a return to the fundamentals of the industrial age—the basis for Germany's Economic Miracle during the postwar period, when Trockel was a child. The machine holds seven rows of eight paint brushes suspended above the ground; when in motion, they dip into China tusche ink, then apply themselves to the Japanese paper below. The significance of the brushes, made with the hair of fifty-six different contemporary artists, translates indexically to the images they produce. Here Trockel mocks the art world by making a fetish of the artist-as-celebrity. Each artist's name is imprinted on the handle of his or her brush, reconfiguring the essence of identity and offering another whimsical and ironic twist to the conventions of art-making.

The paintings produced are strictly nonobjective, bearing no conceivable relation to the personalities of their artistic "authors" and reaffirming Trockel's intention to challenge the art world, and the individual, on the necessity of traditionally ascribed roles and identities. At first the vertical lines seem deceptively simple. Yet when one comes to understand the origin of the marks on paper, that simplicity may be transformed into the marks of celebrity, the relic, or the fetish. Does this give more or less agency and meaning to the drawing? Evidence of the painting machine's actions, the seven drawings are icons of icons, remnants of remnants, and echoes of impressions. The sidewalk in front of the Chinese Theater in Hollywood holds the impressions from the hands and feet of entertainment celebrities in a repetitive and similarly "concrete" way. *56 Brush Strokes* also encapsulates the "impressions" of the celebrities of contemporary art, for exhibition in the environments where their art might be shown, celebrated, and consumed.

These are not formal portraits of artists, nor are they thumbprints or body casts. Instead, they are indexical images produced by the artists' own corporeality, applied to paper by a mechanical device. The machine is engineered, designed, and built by Trockel, but the drawings themselves are untouched directly by any artist's hand. Thus Trockel comments on her contemporaries from a mechanical distance.

—Isabelle Sobin

---

1. Holland Cotter, "Rosemarie Trockel," *Arts Magazine* 65 (January 1989): 81.

Rosemarie Trockel

Ich kannte mich nicht aus
(I am Stumped or I Don't Know About Myself), 1988

Wood, steel, wool, black and white photograph

46 x 20 x 20 in.

Private Collection, Hastings-on-Hudson and Karlsruhe (Courtesy Barbara Gladstone Gallery, New York, New York)

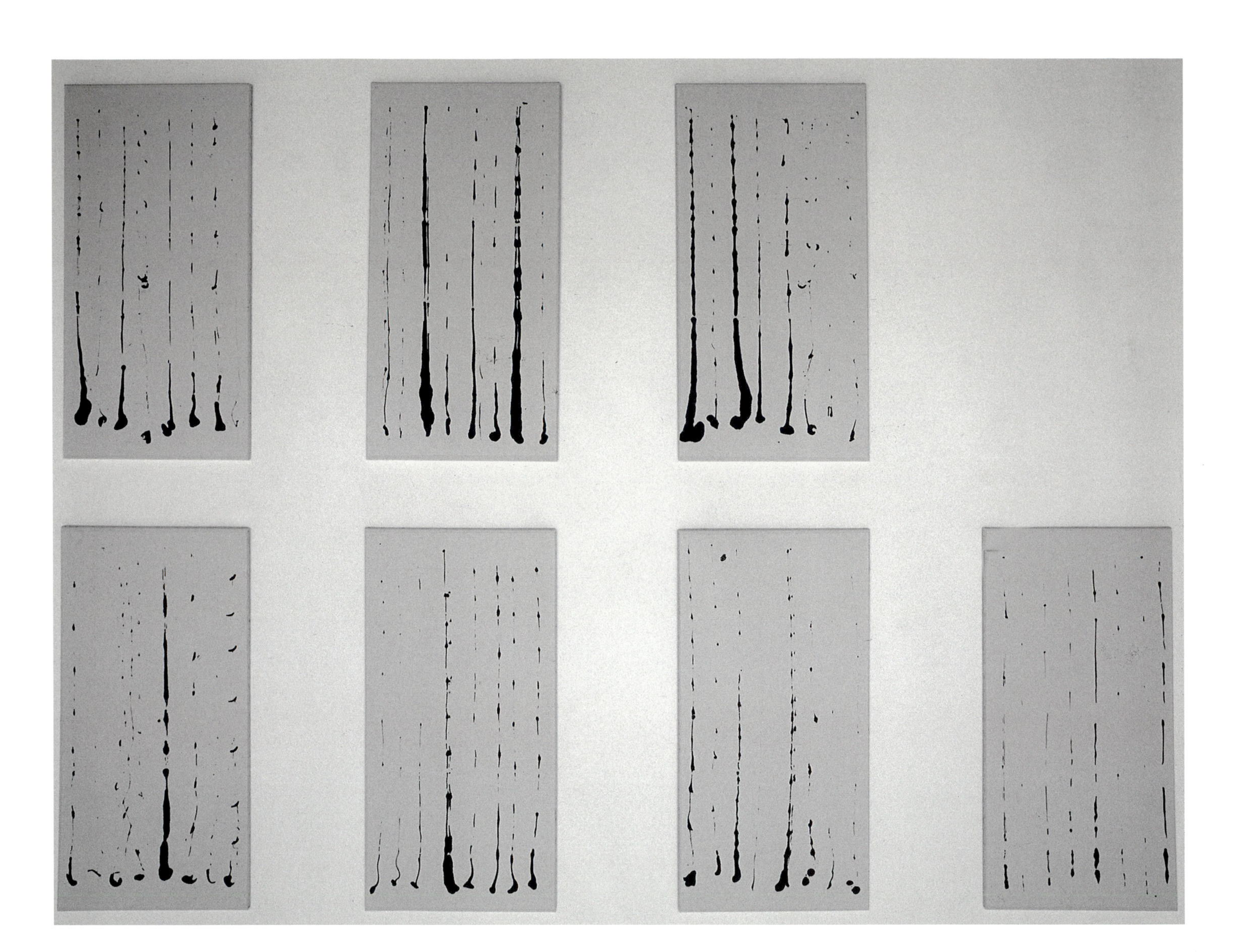

Rosemarie Trockel

56 Brush Strokes, 1990

Seven drawings; China tusche on Japanese paper on canvas, each 55 1/8 x 27 1/2 in.

Monika Sprüth Galerie, Cologne, Germany

EXHIBITION CHECKLIST

1. Angela Bulloch
   Pushmepullme Drawing Machine, 1991
   Steel armature, electric motor, felt-tip marker, chair, and foot switch
   120 x 360 in.
   Private collection, courtesy Deitch Projects, New York
   Page 33

2. Lawrence Gipe
   Painting #4: Themes for a fin-de-siécle, 1989
   Oil on wood panel
   90 x 90 in.
   Private collection
   Page 22

3. Lawrence Gipe
   Triptych from the Krupp Project, 1990
   Oil on wood panel
   Central panel—90 x 70 in.; side panels—90 x 80 in. each
   Los Angeles County Museum of Art, Gift of Barbara Schwartz in Memory of Eugene M. Schwartz
   Pages 36–37

4. Rebecca Horn
   Kleine Malschule (The Little Painting School), 1988
   Metal construction, ladles, brushes, motor, wood stretchers, and pigment
   116 x 17 x 43 in.
   Private collection Switzerland, courtesy London Projects
   Page 39

5. Liz Larner
   Wall Scratcher, 1988
   Anodized aluminum, spring steel, 12-volt motor and battery
   47 x 18 1/8 x 12 1/4 in.
   Collection: Ira and Lori Young, West Vancouver, Canada
   Page 43

6. Liz Larner
   Machine Drawings, 1996–7
   Series of nine mixed-media Iris print collages
   20 x 24 in. each
   Courtesy of the artist
   Pages 40, 58–61

7. Robert Moskowitz
   Stack, 1979
   Oil on canvas
   108 x 34 1/2 in.
   Private collection
   Page 45

8. Donald Sultan
   Plant, May 29, 1985
   Latex, tar, and fabric on vinyl tile over masonite
   96 x 96 in.
   Hirshhorn Museum and Sculpture Garden, Smithsonian Institution. Thomas M. Evans, Jerome L. Greene, Joseph H. Hirshhorn, and Sydney and Frances Lewis Purchase Fund, 1985
   Page 46

9. Donald Sultan
   Polish Landscape II, January 5, 1990
   Latex and tar on vinyl tile over masonite
   96 x 96 in.
   The Eli Broad Family Foundation, Santa Monica
   Page 48

10. Mark Tansey
    Diptych:
    Hub (left panel), 1992
    Graphite on paper
    9 1/4 x 7 in.
    Discursive Formation (right panel), 1992
    Toner on paper
    9 1/4 x 7 1/4 in.
    The Curt Marcus Gallery, New York
    Pages 62–63

11. Mark Tansey
    The Raw and the Framed, 1992
    Toner on paper
    6 x 10 in.
    The Eli and Edythe L. Broad Collection, Los Angeles
    Page 6

12. Mark Tansey
    The Matrix, 1993
    Oil on canvas
    104 x 74 in.
    The Eli and Edythe L. Broad Collection, Los Angeles
    Page 50

13. Rosemarie Trockel
    Untitled (Woolmark), 1986
    Machine-knitted wool
    86 1/2 x 16 in.
    Marjory Jacobson and Marshall Smith, Boston
    Page 52

14. Rosemarie Trockel
    Ich kannte mich nicht aus (I Am Stumped, or I Don't Know About Myself), 1988
    Wood, steel, wool, black-and-white photograph
    46 x 20 x 20 in.
    Private collection, Hastings-on-Hudson and Karlsruhe (courtesy Barbara Gladstone Gallery, New York)
    Page 55

15. Rosemarie Trockel
    56 Brush Strokes, 1990
    Seven drawings: China tusche on Japanese paper on canvas
    55 1/8 x 27 1/2 in. each
    Monika Sprüth Galerie, Cologne, Germany
    Page 56

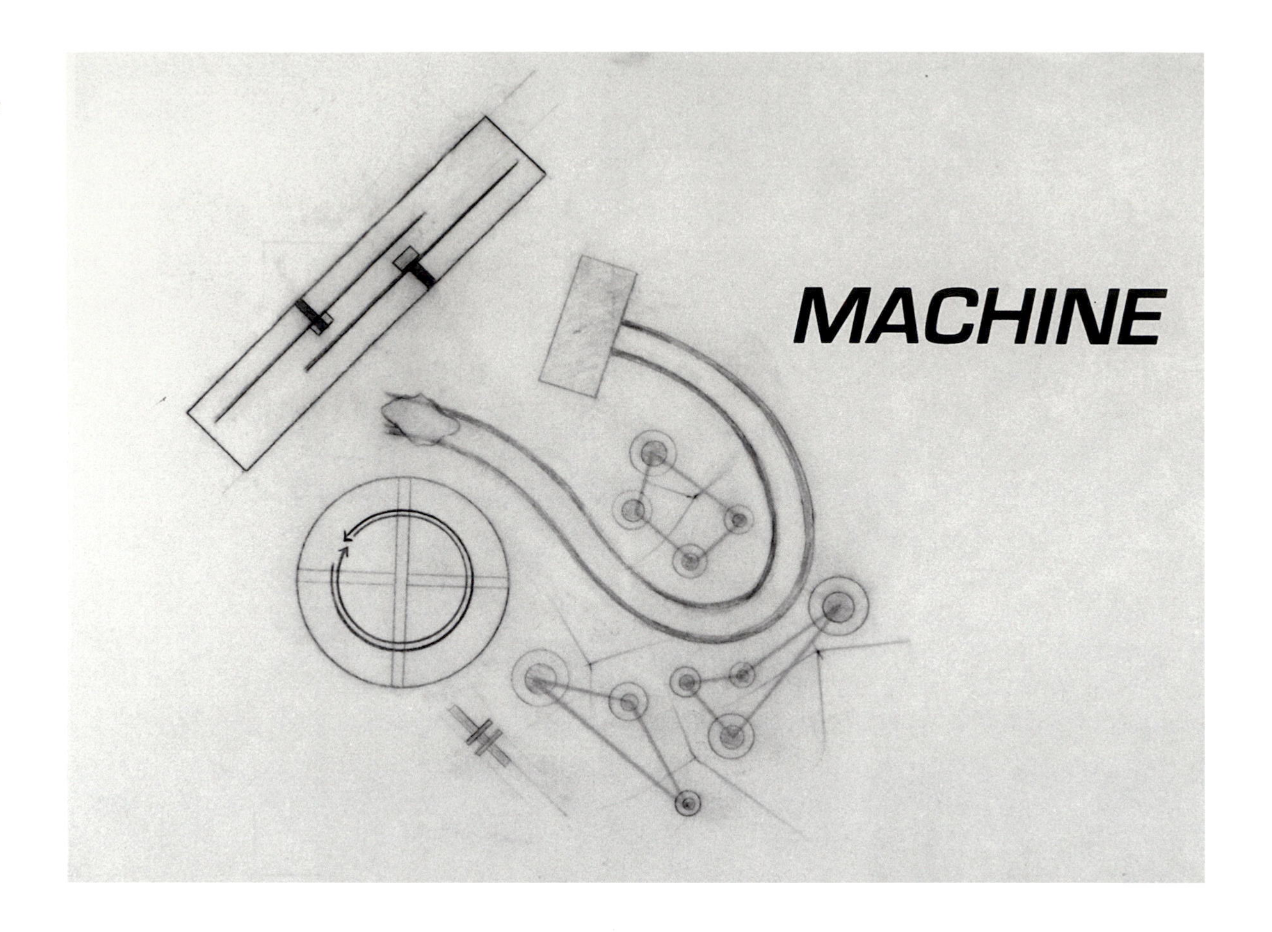

Liz Larner

Title Page from *Machine Drawings*, 1996–7

Mixed-media Iris print collage

20 x 24 in.

Courtesy of the artist

The machine's purpose is to act on itself. The machine produces actions.
The machine's actions are always induced through the relationship of the machine to the sensing field. This relationship is constituted by movement and heat usually in the form of a person or people within the machine's sensing field. The action of the machine is not to break itself down, on the contrary, the machine is engineered to work itself without interruption by mechanical or technological error or failure.
The machine is connected to the experience of encountering an invisible but extremely sensitive curved shape, the sensing field. The machine is a diversion from realizing how it is engaged. The invisible sensorial area that surrounds the machine is a curved shape (this shape may not always be the same but will always be curved) made by electronic eyes -- heat and motion sensors. As one enters this area or broaches this curved shape the machine is turned on. As long as there is movement and heat within the field the machine is on. The only way to stop the machine, or turn it off, is to leave and remain outside the field or to stay still within it.
The machine consists of multiple actions. Some of the actions, like hydraulic pistons and frame, this as that on curved track , are companion mechanisms which are different actions and motions that combine to construe one or more complicated actions. Other components of machine are the enabling lace, which is that group of components which function to mesh and translate the various trajectories and actions that make up the machine. These components would be the computer and fuzzy logic, the flywheel, the cvt, the hydraulic lines, the power sources, actuators, motors and so on.
Other actions of the machine such as squish, fans and feeble are also linked to the enabling but are seen as more self contained motions than the actions of the companion mechanism and the funtion of the enabling lace. As the machine is engaged an action commences. As heat and motion continue to occur in the sensing field the machine's actions will sequentially engage until all mechanisms are working simultaneously, if movement is still occurring in the field. If movement in the field stops the machine will also stop.
The machine's actions are not random but a sequenced set of events that occur in relation to the intensity of heat and motion in the field. This relation, between heat and motion in the field and sequencing and speed of machine's actions, is one of translation (fuzzy logic) as opposed to mimickery.

Liz Larner

Overall from *Machine Drawings*, 1996–7

Mixed-media Iris print collage

20 x 24 in.

Courtesy of the artist

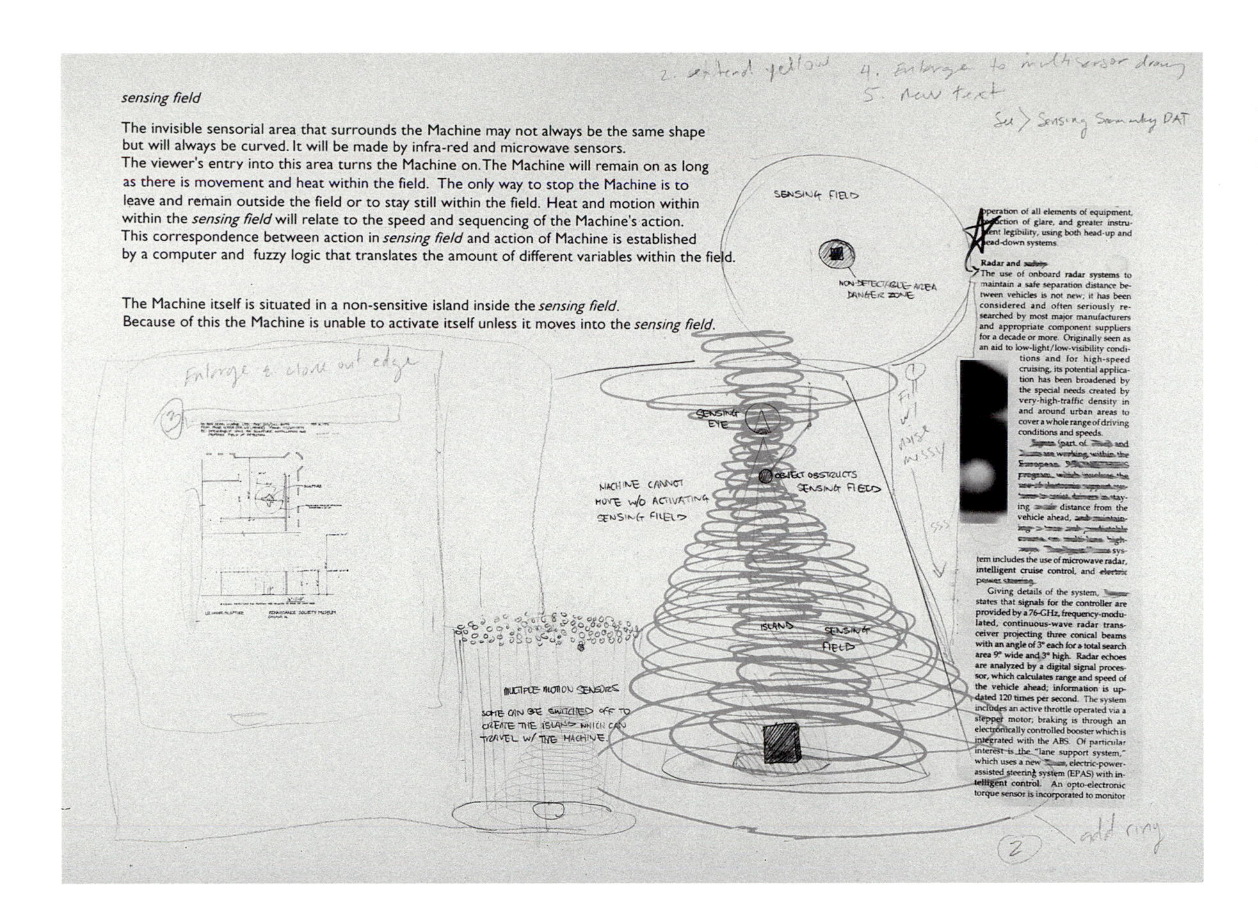

Liz Larner

Sensing Field from *Machine Drawings*, 1996–7

Mixed-media Iris print collage

20 x 24 in.

Courtesy of the artist

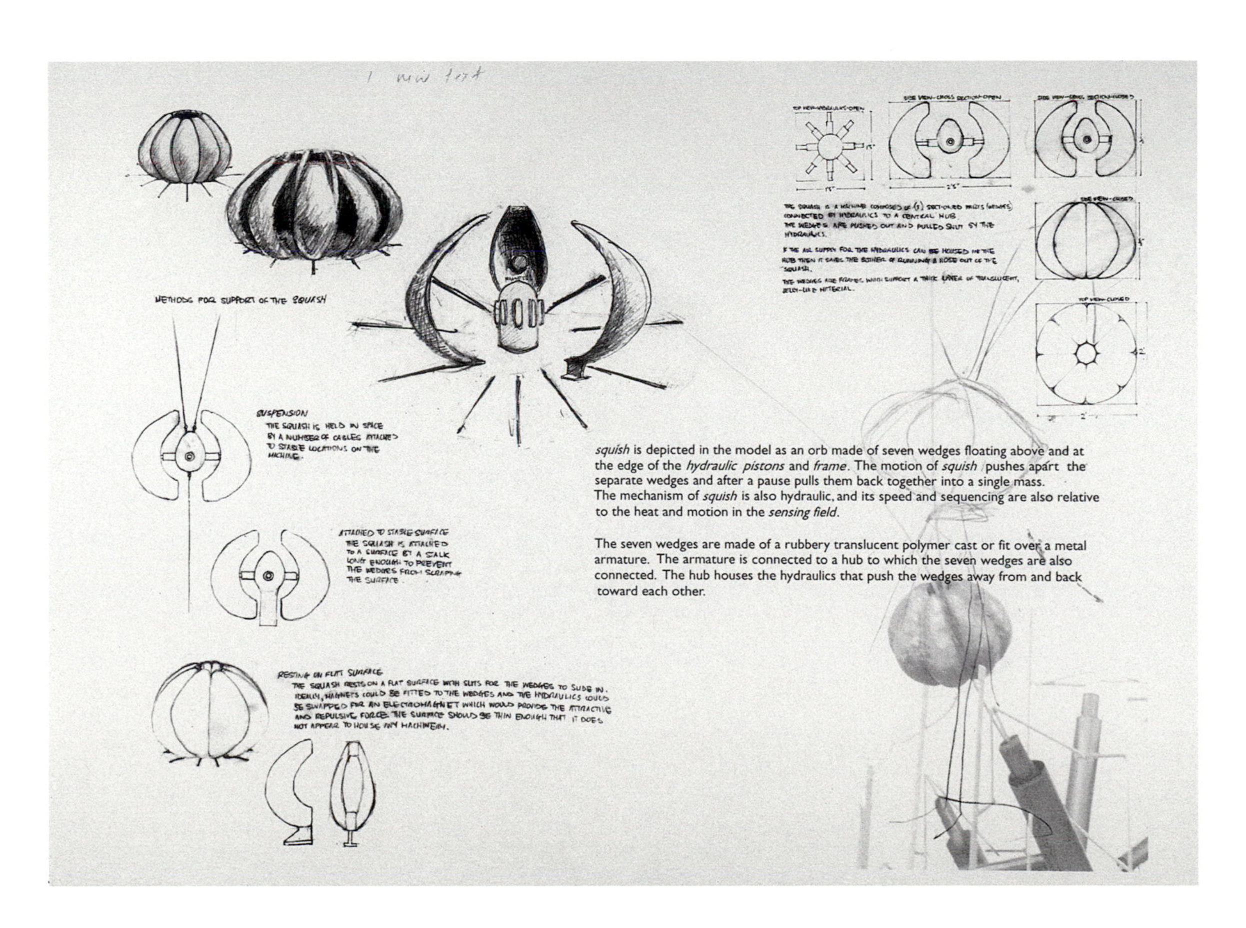

Liz Larner

Squish from *Machine Drawings*, 1996–7

Mixed-media Iris print collage

20 x 24 in.

Courtesy of the artist

Liz Larner

Hydraulic Pistons and Frame from *Machine Drawings*, 1996–7

Mixed-media Iris print collage

20 x 24 in.

Courtesy of the artist

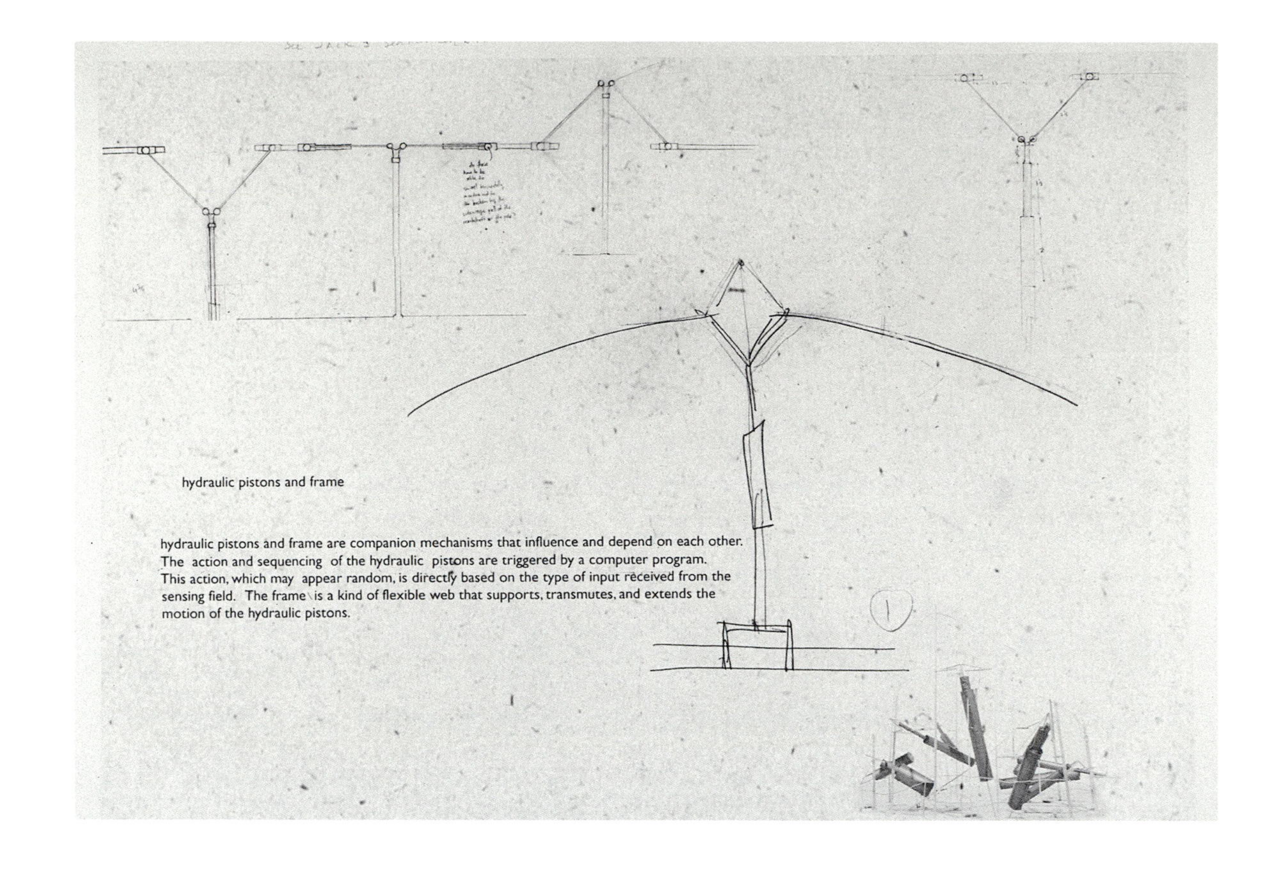

hydraulic pistons and frame
hydraulic pistons and frame are companion mechanisms that influence and depend on each other. The action and sequencing of the hydraulic pistons are triggered by a computer program. This action, which may appear random, is directly based on the type of input received from the sensing field. The frame is a kind of flexible web that supports, transmutes, and extends the motion of the hydraulic pistons.

Liz Larner

Hydraulic Pistons from *Machine Drawings*, 1996–7

Mixed-media Iris print collage

20 x 24 in.

Courtesy of the artist

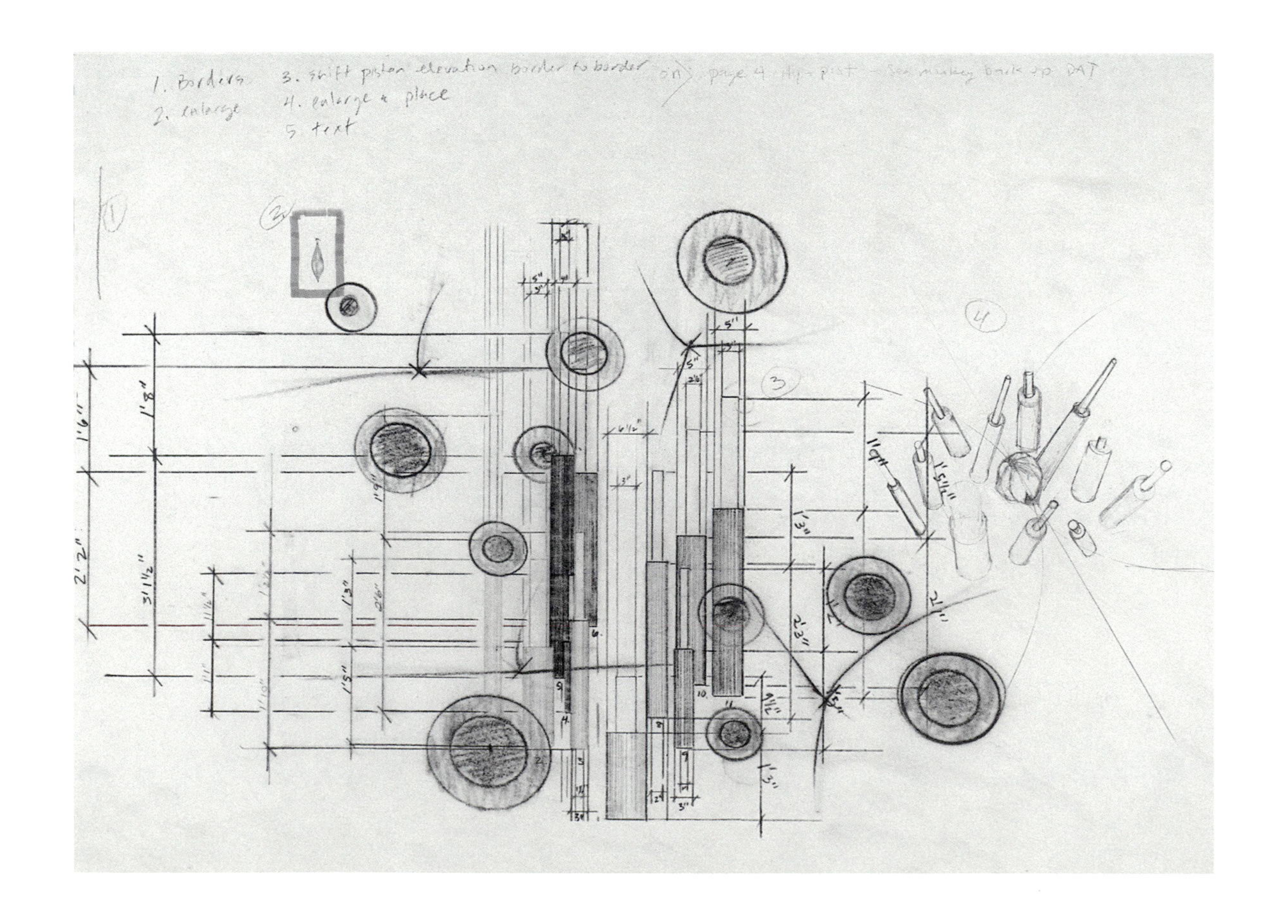

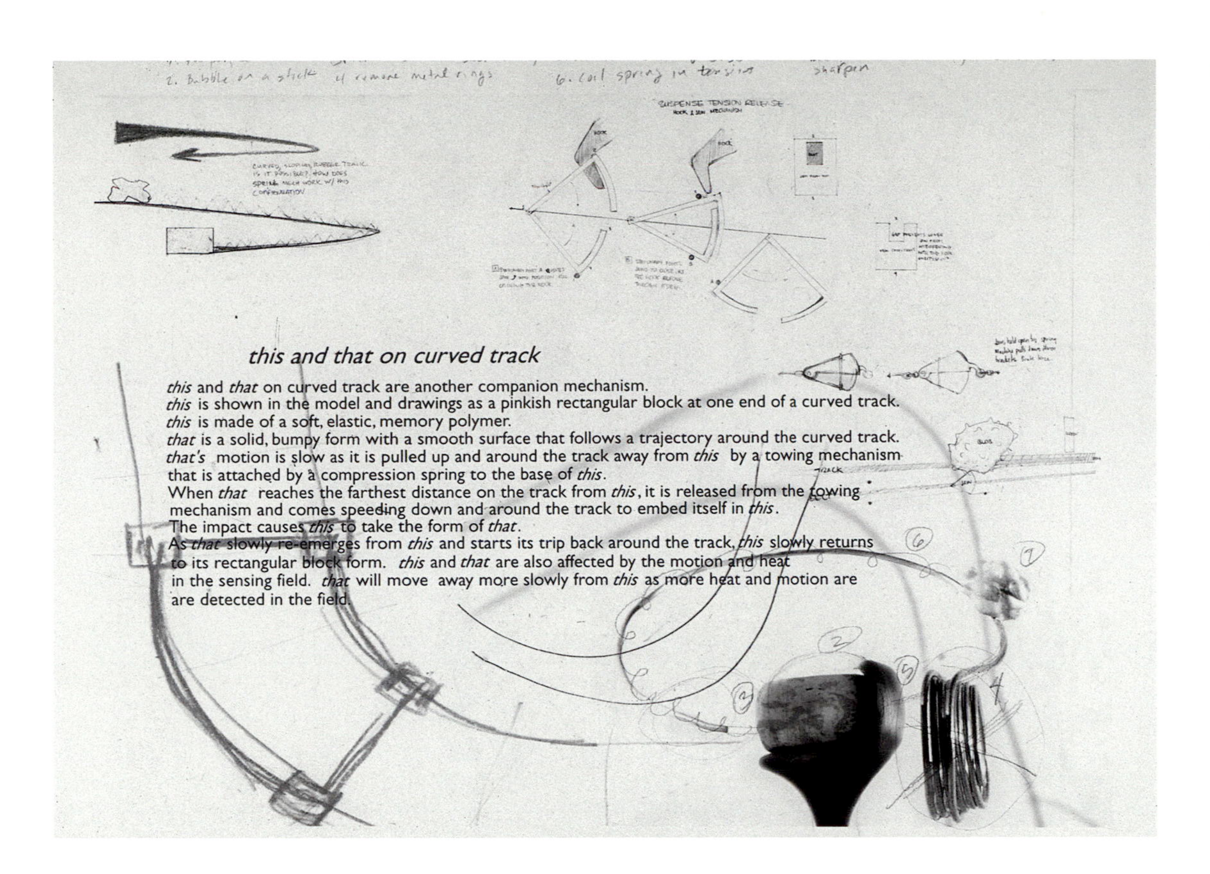

SUSPENSE TENSION RELEASE
this and that on curved track
this and that on curved track are another companion mechanism.
this is shown in the model and drawings as a pinkish rectangular block at one end of a curved track.
this is made of a soft, elastic, memory polymer.
that is a solid, bumpy form with a smooth surface that follows a trajectory around the curved track.
that's motion is slow as it is pulled up and around the track away from this by a towing mechanism
that is attached by a compression spring to the base of this.
When that reaches the farthest distance on the track from this, it is released from the towing
mechanism and comes speeding down and around the track to embed itself in this.
The impact causes this to take the form of that.
As that slowly re-emerges from this and starts its trip back around the track, this slowly returns
to its rectangular block form. this and that are also affected by the motion and heat
in the sensing field. that will move away more slowly from this as more heat and motion are
are detected in the field.

Liz Larner

This and That from *Machine Drawings,* 1996–7

Mixed-media Iris print collage

20 x 24 in.

Courtesy of the artist

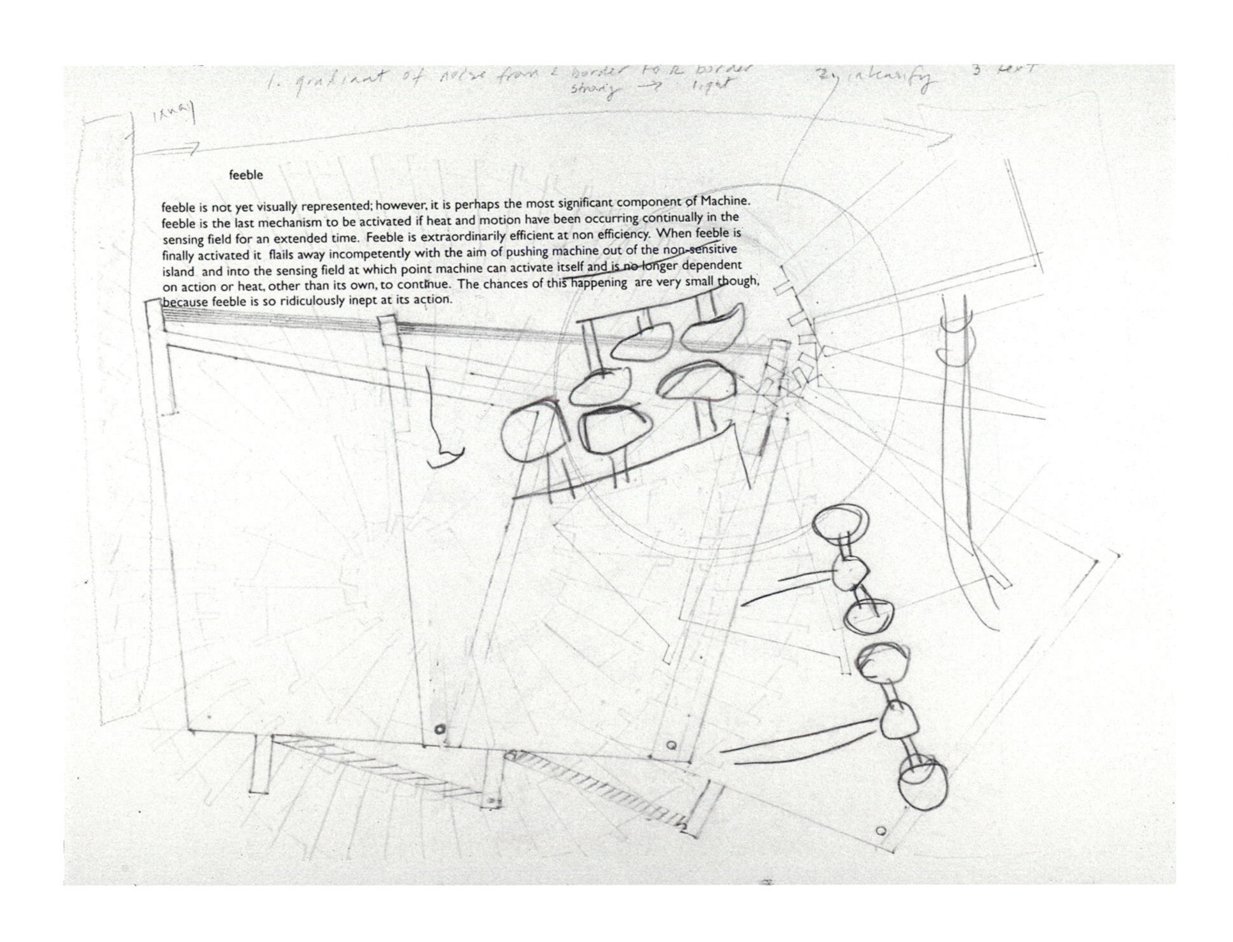

feeble is not yet visually represented; however, it is perhaps the most significant component of Machine.
feeble is the last mechanism to be activated if heat and motion have been occurring continually in the
sensing field for an extended time. Feeble is extraordinarily efficient at non efficiency. When feeble is
finally activated it flails away incompetently with the aim of pushing machine out of the non-sensitive
island and into the sensing field at which point machine can activate itself and is no longer dependent
on action or heat, other than its own, to continue. The chances of this happening are very small though,
because feeble is so ridiculously inept at its action.

Liz Larner

Feeble from *Machine Drawings,* 1996–7

Mixed-media Iris print collage

20 x 24 in.

Courtesy of the artist

Mark Tansey

Diptych: *Hub* (left panel), 1992

Graphite on paper

9 1/4 x 7 in.

The Curt Marcus Gallery, New York

Mark Tansey

Diptych: *Discursive Formation* (right panel), 1992

Toner on paper

9 1/4 x 7 1/4 in.

The Curt Marcus Gallery, New York

Boston University

President: Jon Westling

Provost: Dennis D. Berkey

College of Arts and Sciences

Dean: Dennis D. Berkey

Chairman, Art History Department: Patricia Hills

School for the Arts

Dean: Bruce MacCombie

Director, Visual Arts: Hugh O'Donnell

Boston University Art Gallery

855 Commonwealth Avenue

Boston, Massachusetts 02215

617/353-4672

Director: Kim Sichel

Assistant Director: John R. Stomberg

Security: Evelyn Cohen

Gallery Assistants:

Gina Borg, Katie Delmez, Rachelle A. Dermer, Michéle Furst, Lawrence Hyman, Katrina A. Jones, Alice Kim, Mary Matson, Stacey McCarroll, Emre Pala, Isabelle Sobin, and Andrew Tosiello

Catalogue Designer: Diane Sawyer

Copy Editor: Joseph Crawford

Photographic Credits

All photographs are courtesy of the individual artists, except figures 1–3 (q.v.) and as noted here—Los Angeles County Museum of Art (pages 36–37); London Projects, London (page 39); Hirshhorn Museum and Sculpture Garden, Smithsonian Institution (page 46); Curt Marcus Gallery, New York (pages 62–63); The Eli and Edythe L. Broad Collection, Los Angeles (pages 6, 50); Barbara Gladstone Gallery, New York (page 55).